SpringerBriefs in Philosophy

SpringerBriefs present concise summaries of cutting-edge research and practical applications across a wide spectrum of fields. Featuring compact volumes of 50 to 125 pages, the series covers a range of content from professional to academic. Typical topics might include:

- A timely report of state-of-the art analytical techniques
- A bridge between new research results, as published in journal articles, and a contextual literature review
- A snapshot of a hot or emerging topic
- An in-depth case study or clinical example
- A presentation of core concepts that students must understand in order to make independent contributions

SpringerBriefs in Philosophy cover a broad range of philosophical fields including: Philosophy of Science, Logic, Non-Western Thinking and Western Philosophy. We also consider biographies, full or partial, of key thinkers and pioneers.

SpringerBriefs are characterized by fast, global electronic dissemination, standard publishing contracts, standardized manuscript preparation and formatting guidelines, and expedited production schedules. Both solicited and unsolicited manuscripts are considered for publication in the SpringerBriefs in Philosophy series. Potential authors are warmly invited to complete and submit the Briefs Author Proposal form. All projects will be submitted to editorial review by external advisors.

SpringerBriefs are characterized by expedited production schedules with the aim for publication 8 to 12 weeks after acceptance and fast, global electronic dissemination through our online platform SpringerLink. The standard concise author contracts guarantee that

- an individual ISBN is assigned to each manuscript
- each manuscript is copyrighted in the name of the author
- the author retains the right to post the pre-publication version on his/her website or that of his/her institution.

David Ellerman

Partitions, Objective Indefiniteness, and Quantum Reality

The Objective Indefiniteness Interpretation
of Quantum Mechanics

 Springer

David Ellerman ⓘ
University of Ljubljana
Ljubljana, Slovenia

ISSN 2211-4548   ISSN 2211-4556 (electronic)
SpringerBriefs in Philosophy
ISBN 978-3-031-61785-0   ISBN 978-3-031-61786-7 (eBook)
https://doi.org/10.1007/978-3-031-61786-7

This Springer imprint is published by the registered company Springer Nature Switzerland AG
The registered company address is: Gewerbestrasse 11, 6330 Cham, Switzerland

If disposing of this product, please recycle the paper.

To the memory of Abner Shimony
—Physicist, Philosopher, and Friend.

Preface

My main formation as a mathematician was in collaboration in the early 1980s with Gian-Carlo Rota of MIT. We wrote a joint paper that took my Erdös number from infinity down to 3. But most of my employment in later years was in Economics, not mathematics.

A big shock came in 1999 just before my retirement when Rota died relatively young at age 66. Some of us who had worked with him wanted to pick up some of his unfinished strands to further develop. One of those strands was a 1996 paper by Rota and some colleagues entitled: "Logic of Commuting Equivalence Relations." Rota was well aware of the category-theoretic duality between subsets and partitions (or equivalence relations). Since ordinary logic starts with the Boolean logic of subsets (usually presented in the special case of "propositional logic"), Rota had the idea of developing a logic of that dual concept, a logic of partitions or equivalence relations. But the use of the word "logic" in that 1996 paper was an overstatement since there was no known implication operation on equivalence relations or partitions, only the lattice operations of join and meet known in the nineteenth century (e.g., Dedekind and Schröder).

In fact, no new operations of partitions, such as the implication operation, were developed throughout the twentieth century. Eventually, perhaps with a little luck, I was able to define the implication operation on partitions. Soon it became clear that there were, in fact, two algorithms that could be used to define all the Boolean operations on partitions. That was the beginning of partition logic developed in my recent book, *The Logic of Partitions: With Two Major Applications*.

The first major application was again foreshadowed by Rota who emphasized the analogy: $\frac{\text{Probability}}{\text{Subsets}} \approx \frac{\text{Information}}{\text{Partitions}}$. Boole logically developed finite probability theory starting as the quantitative notion of subsets, e.g., the probability of getting a subset S of a finite equiprobable sample space U is just the normalized number of elements: $\Pr(S) = \frac{|S|}{|U|}$. Hence, using Rota's equivalence, the *logical* notion of information should start with the quantitative notion of the "size" of a partition. The duality of subsets and partitions reveals an underlying duality between elements of subsets and the ordered pairs called the *distinctions* or *dits* of the partition, i.e., ordered pairs in different blocks of a partition.

Then we can finally answer the question raised by Rota's analogy; the quantitative notion of a partition is the size of its set of distinctions, its ditset. The basic logical notion of information in a partition is just the normalized size of its ditset, so the initial definition of *logical entropy* for $\pi = \{B_1, ..., B_m\}$ is

$$h(\pi) = \frac{|\text{dit}(\pi)|}{|U \times U|} = \frac{\left|U \times U - \cup_j(B_j \times B_j)\right|}{|U \times U|} = 1 - \sum_j \left(\frac{|B_j|}{|U|}\right)^2 = 1 - \sum_j \text{Pr}(B_j)^2 =$$

$$\sum_{j \neq k} \text{Pr}(B_j)\text{Pr}(B_k)$$

with equiprobable points in U. If there is a general probability distribution $p = (p_1, ..., p_n)$ on the points of U, then the logical entropy of π is just the value of product probability measure $p \times p$ on the ditset $\text{dit}(\pi) \subseteq U \times U$. The interpretations of $\text{Pr}(S)$ and $h(\pi)$ are thus also analogous. One random draw from U gets an element of S with the probability $\text{Pr}(S)$, and two random draws from U get a distinction of π with the probability $h(\pi)$. This also (finally) gives a logical definition of information, i.e., information as distinctions.

The quantum version, quantum logical entropy, is one of the topics briefly developed in this book. It has the analogous interpretation, i.e., the quantum logical entropy resulting from a projective measurement of an observable on a state is the probability that in two independent measurements of the same observable on an identically prepared state, different eigenvalues are obtained.

Hence the first major application of partition logic is simply its quantitative version as a logical definition of information analogous to the way Boole approached probability theory as a quantitative version of subset logic. That supplies a much-needed *logical* foundation for information theory (classical and quantum) developed in my 2021 book *New Foundations for Information Theory: Logical Entropy and Shannon Entropy.*

The second major application and the topic of this book is the century-old problem of understanding the reality that quantum mechanics (QM) describes so well. QM was consolidated in the mid-1920s but, over the last century, there has been no agreement on the nature of reality at the quantum level. New so-called "interpretations" are continually being created without any noticeable convergence. Otherwise sane physicists are driven to rather bizarre ideas, e.g., the many-worlds interpretation, when confronted with the "paradoxes" of quantum theory. It is in this intellectual "demolition derby" of quantum interpretations where partition logic and logical entropy offer a new approach to corroborate (in a suitably reformulated manner) an interpretation already promoted by Werner Heisenberg and Abner Shimony, among others.

This new approach first "cuts at the joint" between the mathematics and the physics of quantum mechanics. The mathematics is quite distinctive and different from the mathematical framework of classical physics. The new approach asks:

Where does the distinctive mathematics of QM come from?

The answer is that the math of QM is the vector space or, particularly, Hilbert space version of the mathematics of partitions. The argument is based, in part, on using a semi-algorithmic procedure, herein called the *Yoga of Linearization* (part of mathematical folklore), to build a translation dictionary between set-level partition math and Hilbert space QM math.

For instance, here is a set-level construction in partition math whose quantum math version is Dirac's notion of a Complete Set of Commuting Observables (CSCO).

Partition-math version: A set $f, ..., g : U \to \mathbb{R}$ of real-valued numerical attributes on a set $U = \{u_1, ..., u_n\}$ is said to be *complete* (a Complete Set of Compatible Attributes or CSCA) if the join (non-empty intersections of the blocks) of their inverse-image partitions is the partition with all blocks of cardinality one. Then each element u_i of U is uniquely characterized by its ordered set of values for $f, ..., g$.

Quantum-math version: A set $F, ..., G$ of commuting observables on a Hilbert space V is said to be *complete* (a Complete Set of Commuting Observables or CSCO) if the join (non-zero intersections of the eigenspaces) of their direct-sum decompositions (DSDs) of eigenspaces is the direct-sum decomposition with all subspaces of dimension one. Then each eigenvector v_i in the set of simultaneous eigenvectors spanning V is uniquely characterized by its ordered set of eigenvalues for $F, ..., G$.

Each version is essentially a word-for-word translation using the following translation dictionary. Many of the tables in the book are additions to the translation dictionary to show how the QM math (right side of the table) is the vector space, and particularly Hilbert space version of the math of partitions (left side of the table).

Partition math	Quantum math
Real-valued attributes on a set	Observable ops. on a Hilbert space
Attributes defined on same U	Observables that are commuting
Domain U of the attributes	Basis simult. eigenvectors of comm. ops.
Inv.-image partition of attribute	DSD of eigenspaces of an observable
Join of inverse-image partitions	Join of DSDs of commuting observables
Cardinality of subset in partition	Dimension of subspace in DSD
Partition blocks of cardinality 1	DSD subspaces all of dimension 1
Values of attributes on $u_i \in U$	Eigenvalues of simultaneous eigenvectors
$u_i \in U$ given by attrib. values	Eigenvectors given by eigenvalues

Illustration of QM math being Hilbert space version of partition math

The next step in the argument is to ask:

What basic concepts are represented at the logical level by partitions?

The answer is the concepts described in various vocabularies as indistinctions versus distinctions, indefiniteness versus definiteness, indistinguishability versus distinguishability, equivalence versus inequivalence, or difference versus identity. These pairs of concepts, described by different words in different contexts, might be referred to as the "identity & difference concepts." The *logic* of those identity & difference concepts is the logic of partitions (or equivalence relations) on a set.

Then to interpret QM, we ask:

What is the essential non-classical concept in QM?

The answer is the notion of superposition (entanglement being a particularly vexing special case). But that non-classical notion has been emphasized from the beginning (e.g., by Dirac) so,

Why has there been so little progress in understanding the reality behind the notion of superposition?

The answer lies in the mathematics of QM itself. A Hilbert space is a vector space over the complex numbers $\mathbb{C}$, and the complex numbers are the natural mathematics to describe waves, i.e., the polar representation of a complex number is an amplitude and phase of a wave. In fact, QM was often called "wave mechanics" and the "wave function" is a commonly used mathematical tool to represent the quantum state. Hence, superposition has usually been interpreted simply like the addition of waves—just as water waves might add and interfere with each other in the classroom ripple tank model of the double slit experiment. But after a century of looking, no physical reality has been found for the wave functions—much to the dismay of Erwin Schrödinger who invented the wave equation bearing his name. The wave functions were "probability waves" which are not physical entities at all.

Indeed, the math of QM is formulated using the complex numbers for reasons that have nothing to do with waves, namely that the complex numbers are the algebraically complete extension of the real numbers so that the real-valued quantum observables will then have a complete set of eigenvectors. The whole wave interpretation of QM math was mistakenly giving an ontological importance to the wave-like computational artifacts present in any vector space over the complex numbers.

How to escape this conundrum that the wave-like math is not reflected in quantum level ontology?

What is needed is a *totally different interpretation of superposition* (than the mathematically correct but ontologically misleading addition of vectors that can be interpreted as waves). And *that* new interpretation is supplied by the mathematics of partitions. At the simple logical level, a partition is made up of blocks of elements of the underlying set. Each block is an equivalence class that says, according to this partition, these elements in the block are equated, blobbed, blurred, and cohered together with no distinctions between them—since the distinctions are between different blocks. Thus, the blocks (or equivalence classes) with two or more elements are the logical version of a superposition of eigenstates in QM math. The block is indefinite or indistinct on the differences between elements (or eigenstates)—and definite on commonalities. The elements in the set (or eigenstates in the QM version) represent (not different particles but) different states of a particle that are equated, blobbed together, or cohered together in the superposition. That is the *non-wave indefiniteness reinterpretation of superposition* that corroborates an interpretation of QM proposed by Heisenberg, Shimony, and others. The key idea of this version of superposition is that a particle can be in an objectively indefinite state like a particle in a superposition state of "here" and "there", i.e., it is "not definitely here and not definitely there, but definitely not anywhere else."

Quantum theorists constantly use the wave-math without finding any physical waves, and they (mostly) recognize the reality of indefinite states. That is, when the quantum state is a superposition in the basis of the observable being measured, then it is widely recognized that the quantum state does not have a definite value before the measurement which causes the quantum jump into a definite state. And the set-level version of (projective) measurement is just the partition join operation from partition logic. Heisenberg, Shimony, and others then extrapolate that notion of an indefinite state to the whole of quantum-level reality. The quantum world is Indefinite World. And the set-level mathematics to represent definiteness versus indefiniteness is the math of partitions with the math of QM being the Hilbert space version of that partition math. That new partition-math approach to the Heisenberg-Shimony Objective Indefiniteness Interpretation of QM is the topic of this book.

In addition to the influence of Gian-Carlo Rota and Abner Shimony (my undergraduate advisor at MIT '65), I would like to acknowledge the assistance of the late Larry Harper, Brian Linard, John DePillis, and Tom Payne who were the members of the "Schmooze Group" of retired professors studying quantum mechanics at the University of California at Riverside.

Ljubljana, Slovenia   David Ellerman
March 2024

Competing Interests The author has no conflicts of interest to declare.

Contents

Chapter 1
Introduction: Partitions and Quantum Reality

Philosophy is written in this grand book the universe, which stands continually open to our gaze. But the book cannot be understood unless one first learns to comprehend the language and to read the alphabet in which it is composed. It is written in the language of mathematics, and its characters are triangles, circles, and other geometric figures, without which it is humanly impossible to understand a single word of it; without these, one wanders about in a dark labyrinth.

Galileo

Abstract This new approach to understanding and interpreting quantum mechanics (QM) is based on the development of the notion of a partition on a set (or, equivalently, an equivalence relation on a set or a quotient set). The notion of a partition is not some ad hoc notion designed to make yet another interpretation of quantum mechanics. It is as fundamental a notion as that of a subset. But the math (and logic) of subsets was developed long before the corresponding partition math. There is a recent sequence of developments that constitute a revival or renaissance of partitions outside of the standard treatment in enumerative combinatorics. Standard logic is the Boolean logic of subsets (propositional logic being a special case). Subsets and partitions are dual concepts in category theory. The first stage in the revival was the development of the logic of partitions dual to the logic of subsets. Then just as finite Boole-Laplace probability theory starts as the quantitative version of the logic of subsets, so the quantitative version of the logic of partitions was the second stage resulting in the new foundations for information theory based on the notion of logical entropy. The third stage was based on the realization that the distinctive mathematics of QM is the vector (particularly Hilbert) space version of the mathematics of partitions. That is the subject of this book.

Keywords Partitions · Partition logic · Logical entropy · Its and dits duality · Mathematics of quantum mechanics · Wave mechanics · Objective indefiniteness

© The Author(s), under exclusive license to Springer Nature Switzerland AG 2024   1
D. Ellerman, *Partitions, Objective Indefiniteness, and Quantum Reality*,
SpringerBriefs in Philosophy, https://doi.org/10.1007/978-3-031-61786-7_1

1.1   Partition Logic

Much of the literature in the philosophy of quantum mechanics (QM) focuses on alternative ways to change the standard received version of von Neumann/Dirac QM in order to supposedly interpret it realistically, e.g., Bohmian mechanics, spontaneous localization, many-worlds, not to mentioned less-well-known approaches such as the transactional approach. Our approach here is to show how to realistically interpret the standard (von Neumann/Dirac) theory.

The mathematical formalism of quantum mechanics (QM) is quite distinctive in comparison with the mathematics of classical physics. If one feature was to be picked out as the distinctive feature of QM math, it would be the notion of a superposition (and the associated notion of becoming given below), i.e., that the sum of two quantum states is another quantum state.

> Thus, *superposition*, with the attendant riddles of entanglement and reduction, remains *the* central and generic interpretative problem of quantum theory. (Cushing 1988, p. 27)

But where does that distinctive math (not the physics) of QM come from? The question of interpreting QM has been like Galileo's "dark labyrinth" without the focus on that *distinctive* mathematics of QM. We "follow the math" back to its source. Our thesis is that the math of QM is the vector (particularly, Hilbert) space version of the math of partitions. Or, put the other way around, the math of partitions offers a skeletionized (or scalarless) version of QM math. At the level of sets, superposition is modeled by equivalence, i.e., the definite- or eigen-states that are superposed are modeled by elements in the same equivalence class or same block of a partition. This application of partition math to better understand and interpret quantum mechanics is part of a series of developments about partitions–starting with the logic of partitions.

"Logic" is often interpreted to refer to reasoning about propositions. But mathematically, the logic of propositions is only a special case of the Boolean logic of subsets (Boole 1854), i.e., the special case where the universe set U is a one-element set whose two subsets, 1 and $\emptyset$, are taken to represent truth and falsity. The formulas in the Boolean logic of subsets stand for subsets of a universe set U, not propositions. And, again mathematically, the notion of a subset (or more generally, a subobject or 'part') is dual to the notion of a partition or equivalence relation (or quotient object). "The dual notion (obtained by reversing the arrows) of 'part' is the notion of partition." (Lawvere and Rosebrugh 2003, p. 85). But in view of the aforementioned duality, there should also be a dual mathematical logic of partitions where the formulas stand for partitions on a universe set U.

In the nineteenth century, the lattice operations of join and meet were defined for partitions (Richard Dedekind, Ernst Schröder, and others). But to qualify as a "logic" there should at least be an implication operation on partitions–but no such operation was defined throughout the twentieth century. In short, the mathematics of partitions was woefully underdeveloped.

In a 2001 paper commemorating Gian-Carlo Rota (1932–1999),[1] the three authors first note the fundamentality of partitions (or equivalence relations) and then acknowledge the sole operations of join and meet.

> Equivalence relations are so ubiquitous in everyday life that we often forget about their proactive existence. Much is still unknown about equivalence relations. Were this situation remedied, the theory of equivalence relations could initiate a chain reaction generating new insights and discoveries in many fields dependent upon it.
>
> This paper springs from a simple acknowledgment: the only operations on the family of equivalence relations fully studied, understood and deployed are the binary join $\vee$ and meet $\wedge$ operations. (Britz et al. 2001, p. 445)

This anticipated "chain reaction" is the partition revival that started with the development of the logic of partitions. That development required the definition of the logical operation of implication on the lattice of partitions along with several algorithms to define all the Boolean or logical operations on partitions (Ellerman 2010, 2014, 2023).

The logics of subsets and partitions are equally fundamental from the mathematical point of view. The dual logics of subsets and partitions seem to be the only such structures defined solely in terms of the completely unstructured universe sets U; there are no ordering relations, no topologies, and no other structures on U. Since the dual logics assume no other structures on the universe set, they are fundamental in that sense.

1.2   The Partition-Logical Theory of Information

The second wave in the chain reaction or partition revival was the development of the quantitative version of the logic of partitions. In view of the duality between subsets and partitions, the idea is to do for partitions what Boole did for the subsets in the logic of subsets (Boole 1854). By associating with each subset its cardinality, the Boole-Laplace logical notion of probability was initially defined as the normalized size of the subsets. The definition of logical entropy follows the same idea but applied to the dual notion of partitions.

The definition of logical entropy fulfills a program of Gian-Carlo Rota that begins with the idea: "The lattice of partitions plays for information the role that the Boolean algebra of subsets plays for size or probability" (Kung 2009, p. 30). In Rota's Fubini Lectures (and in his lectures at MIT (Rota 1998)), he argued that since partitions are

---

[1] Rota had the idea of developing a logic of equivalence relations or partitions, but with no implication operation, the valid formulas or identities would be the ones that hold for all *lattices* of partitions. And it was shown (Whitman 1946) that lattices of partitions are so versatile that the only identities that hold for lattices of partitions are the identities that hold for all lattices whatsoever. Hence Rota and colleagues (Finberg et al. 1996) developed only the logic (without implication) for a special type of partition or equivalence relation, *commuting* equivalence relations (Dubreil and Dubreil-Jacotin 1939).

dual to subsets, then quantitatively, information is to partitions as probability is to subsets:

$$\frac{\text{Information}}{\text{Partitions}} \approx \frac{\text{Probability}}{\text{Subsets}}$$

Since "Probability is a measure on the Boolean algebra of events" that gives quantitatively the "intuitive idea of the size of a set", we may ask by "analogy" for some measure "which will capture some property that will turn out to be for [partitions] what size is to a set." He then asks:

> How shall we be led to such a property? We have already an inkling of what it should be: it should be a measure of information provided by a random variable. Is there a candidate for the measure of the amount of information? (Rota 2001, p. 67)

It turns out that the notion of a *distinction* of a partition (an ordered pair of elements in different blocks) corresponds to the notion of an element of a subset so "logical entropy" is initially defined as the normalized number of distinctions in a partition (more on this below).

1.3   The Partition Analysis of the Mathematics of Quantum Mechanics

The third wave in the chain reaction or partition revival is the application of partitions to understand the reality that quantum mechanics so successfully describes. After modern quantum mechanics was consolidated almost a century ago, there have been evermore interpretations of it instead of convergence. It is time to ask the simple question: "Where does the distinctive mathematics (not the physics) of quantum mechanics come from?".

The purpose of this book is to show that the mathematics of (standard Dirac-von Neumann non-relativistic) quantum mechanics comes from the set-level partition mathematics of indefiniteness and definiteness linearized to vector spaces, particularly Hilbert spaces. This fact shows the type of reality that QM describes so well; it is characterized by objective indefiniteness and objective probabilities.

> How do definite states arise from indefinite ones? There's a temptation here to think in terms of emergence, but there can be no continuous transition from indefinite to definite states of reality. (Valentini 2011, p. 157)

It is also characterized by the *partition notion of becoming or emergence*, i.e., the *discontinuous* jump from an indefinite state to a more definite state by being in-formed by distinctions. This idea that quantum level reality is characterized by objective indefiniteness is not only embedded in Heisenberg's Indeterminacy Principle but is already hinted at by every quantum theorists who realizes that a superposition of measurement basis eigenstates with different eigenvalues does not have a definite value prior to a state reduction ("measurement"). This indefiniteness in the quantum state has been described by a wide variety of adjectives such as: blurry, unsharp, smudged, blunt, fuzzy, blob-like, dispersed, smeared out, indeterminate, or spread out.

This approach to interpreting QM might be usefully compared to the "reconstruction" approach (Jaeger 2019). In that approach, the development of new and perhaps perspicuous axioms for both the math and physics of QM may give a hint about the underlying ontology. In our approach developed here, the idea is to reconstruct just the new and distinctive math of QM as the Hilbert space version of the logical level math of partitions. That alone suggests the ontology of objective indefiniteness. The physics is then developed through quantization to adapt classical physics to that new mathematical framework—which results in quantum mechanics. The focus on the *math* of QM is evidenced by Planck's constant not being used throughout this book. The program herein of treating the math and physics of QM separately thus seems to be "cutting at the joints." The usual reconstruction programs of treating the math and physics together does not seem to have revealed the ontology underlying QM. The 'secret sauce' is in the QM math; the physics is carried over in an appropriately quantized form from classical physics, e.g., the De Broglie and Einstein formulas.

The development of the logic of partitions, the information theory grounded on logical entropy, and the partition approach to QM are all based on the concepts variously formulated as indistinction versus distinction, indistinguishability versus distinguishability, equivalence versus inequivalence, or identity versus difference. Those are basic *logical* concepts[2]; they provide a fundamental interpretation of QM. They are not yet another set of jury-rigged notions to provide yet another interpretation. It should not be surprising when the most fundamental mathematical concepts are effective in explicating the most fundamental physical theory (Wigner 1967).

1.4   The Classical Versus Quantum Views of Reality

Classical physics exemplified the common-sense idea that reality had fully definite properties. At the logical level, i.e., Boolean subset logic, each element in the Boolean universe set is either definitely in a subset or its complement, i.e., each element either definitely has or does not have a property. Each element is characterized by a full set of properties, a view that might be referred to as "definite all the way down."[3] This was expressed in Immanuel Kant's Principle of Complete Determination (*omnimoda determinatio*).

> Every thing, however, as to its possibility, further stands under the principle of thoroughgoing determination; according to which, among all possible predicates of things, insofar as they are compared with their opposites, one must apply to it. (Kant 1998, B600)

This view of reality was expressed by Leibniz's Principle of the Identity of Indistinguishables. If one could always dig deeper into a fully definite reality to find attributes to distinguish entities, then entities that were completely indistinguishable

---

[2] They might be called "metaphysical" concepts but there is no need for such a vague philosophical term when there is a mathematical *logic* of distinctions and indistinctions, i.e., partition logic (Ellerman 2023).

[3] This metaphor paraphrases the old joke about "turtles all the way down.".

would logically have to be identical. Another principle characteristic of this classical metaphysics of reality was Leibniz's Principle of Continuity (Auletta 2019, p. 7) which could also be expressed as *"Natura non facit saltus"* (Nature does not make jumps) (Leibniz 1996, Bk. IV, Chap. XVI). Yet another principle was Leibniz's Principle of Sufficient Reason "that nothing happens without a reason why it should be so rather than otherwise" (Ariew 2000, Second letter, p. 7).

All of these principles are violated in QM. It is now rather widely accepted that this common-sense fully definite view of reality is not at all compatible with quantum mechanics. The Principle of Identity of Indiscernibles is violated in the theory of indistinguishable particles, the Principle of Continuity is violated by the infamous quantum jumps, and the Principle of Sufficient Reason is violated by the quantum probabilities being objective instead of based on ignorance about the underlying reasons.

If there are only two positions, *here* and *there*, then in classical physics a particle is either definitely *here* or *there*, while in QM, the particle in a superposition state can be objectively "neither definitely here nor there." (Weinberg 1994, p. 75) This is not an epistemic or subjective indefiniteness of location; it is an ontological or objective indefiniteness. The indefiniteness or indistinguishability cannot be resolved by digging deeper with more precision; it is objective. We may look through a microscope or telescope and see a blurred picture and then bring it into a sharp focus, but we know that the viewed reality was always sharp, so the unfocused or indefinite picture was only subjective. But in the view of quantum reality developed here, the indefiniteness is objective. The quantum notion of becoming to turn the indefinite reality into a more definite reality is described at the quantum level as a state reduction.[4] The idea that the ontology underlying QM is characterized by objective indefiniteness should not be a surprise to quantum theorists who note that a superposition eigenstates (in the measurement basis) do not have a definite value prior to the measurement.[5]

The notion of *objective indefiniteness* in QM has been most emphasized by Abner Shimony.[6]

> From these two basic ideas alone – indefiniteness and the superposition principle – it should be clear already that quantum mechanics conflicts sharply with common sense. If the quantum state of a system is a complete description of the system, then a quantity that has an indefinite value in that quantum state is objectively indefinite; its value is not merely unknown by the scientist who seeks to describe the system. ...Classical physics did not conflict with common sense in these fundamental ways. (Shimony 1988, p. 47)

Since our natural common sense view of the world is fully definite (as in classical physics), how can we describe an indefinite reality? Instead of inventing ever more

---

[4] State reductions are often described as "measurements" which is a seriously misleading formulation since it implies a macroscopic human-level apparatus which should have no role in quantum *theory*.

[5] The word "eigen" should be translated or interpreted to mean "definite." An eigenstate is a definite state and an eigenvalue is a definite value.

[6] The vast literature on metaphysical vagueness (e.g., sorites arguments) does not seem helpful to clarify the notion of "quantum indeterminacy" (Lewis 2016).

bizarre "interpretations" of QM, this book looks at the source of the math of QM. The basic mathematical, indeed logical, concept that describes indefiniteness and definiteness is the notion of a *partition (or equivalence relation) on a set*. This book shows that the mathematics of quantum mechanics is the mathematics of partitions linearized to (Hilbert) vector spaces. This substantiates that key analytical concepts in QM are indefiniteness and definiteness, indistinction and distinction, and indistinguishability and distinguishability, e.g., to separate a measurement process from a non-measurement unitary evolution.

1.5 The Basic Logical Operations for Partitions

We first will take a closer look at the math of partitions. Given a universe set $U = \{u_1, ..., u_n\}$, a *partition* $\pi = \{B_1, ..., B_m\}$ is a set of non-empty subsets $B_j \subseteq U$ (for $j = 1, ..., m$) called "blocks" that are disjoint and whose union is U.[7] A *distinction* (or *dit*) of a partition π is an ordered pair of elements $(u_i, u_k) \in U \times U$ in different blocks of π, and dit $(\pi) \subseteq U \times U$ is the set of all distinctions, called the *ditset* of π. An *indistinction* (or *indit*) of π is an ordered pair of elements in the same block of U, and the set of all indistinctions indit (π), called the *indit set*, is the equivalence relation associated with π where the blocks are the equivalence classes. The ditset and indit set of a partition are complements, i.e., they are disjoint and their union is $U \times U$.

Each block $B_j \in \pi$ of a partition should be thought of as being indefinite or indistinct between its elements $u_i, u_k \in B_j$. Partitions naturally arise as the inverse-images $f^{-1} = \left\{ f^{-1}(y) \right\}_{y \in f(U)}$ of functions $f : U \to Y$. In particular, a *numerical attribute* is a function $f : U \to \mathbb{R}$ into some set of values which we can take as the real numbers $\mathbb{R}$. Each block $f^{-1}(r)$ in the partition f^{-1} then represents the constant set of all elements $u_i \in U$ taking the value $f(u_i) = r \in f(U) \subseteq \mathbb{R}$. When the set U is taken as the outcome set or sample space of a finite probability distribution [with equiprobable points or point probabilities $p_i = \Pr(u_i)$], then the numerical attribute is a random variable.

The two main mathematical concepts in QM are observables and states–and partitions prefigure *both* observables and states. As an aid to intuition, these simple concepts at the logical level might be seen as the elementary forms of the more developed mathematical concepts of quantum mechanics as illustrated in Table 1.1 where the partition concepts prefigure quantum observables. These connections will be further developed in the later chapter on the Yoga of Linearization.

In the Boolean logic of subsets, the powerset $\wp(U)$ (set of all subsets of U) forms a lattice where the partial order is set inclusion, the join (least upper bound) and meet (greatest lower bound) are union and intersection respectively, and the top and bottom of the lattice are the universe set U and the empty set $\emptyset$ respectively.

---

[7] Since our purpose is conceptual clarity, not mathematical generality, we will stick to the finite sets and dimensions throughout.

Table 1.1 Partition logical precursors for QM math concepts

Logical concept	QM concept
Block	Eigenspace
Elements in a block	Eigenvectors with same eigenvalue
Numerical attribute	Observable (Hermitian operator)
Elements of U	Basis eigenvectors of an observable
Values of attribute	Eigenvalues of an observable
Partition on a set	Direct-sum decomposition of eigenspaces
Indistinctions	Coherences (density matrix non-zero entries)

In the dual logic of partitions, the set $\Pi (U)$ of partitions on U also forms a lattice where the partial order is refinement. Given another partition $\sigma = \{C_1, ..., C_{m'}\}$ on U, the partition σ is *refined by* π, written, $\sigma \precsim \pi$, if for every block of π, there is a block of σ containing it. Intuitively, the blocks of π can be obtained by chopping up the blocks of σ. If $\pi = \left\{f^{-1}(r)\right\}_{r \in f(U)}$ and $\sigma = \left\{g^{-1}(s)\right\}_{s \in g(U)}$ are the inverse image partitions of random variables $f : U \to \mathbb{R}$ and $g : U \to \mathbb{R}$ respectively, then $\sigma \precsim \pi$ means that the random variable f is *sufficient* for g, i.e., the value of f determines the value of g since each block $f^{-1}(r)$ is contained in some $g^{-1}(s)$.

The *join* $\pi \vee \sigma$ (least upper bound in the refinement ordering) is the partition of U whose blocks are all the non-empty intersections $B_j \cap C_{j'}$. In terms of dit-sets, dit $(\pi \vee \sigma) = $ dit $(\pi) \cup $ dit (σ). The corresponding operation on the associated equivalence relations is their intersection, i.e., indit $(\pi \vee \sigma) = $ indit $(\pi) \cap $ indit (σ).

To form the *meet* $\pi \wedge \sigma$ (greatest lower bound in the refinement ordering), think of two intersecting blocks B_j and $C_{j'}$ as two overlapping blobs of mercury that unify to make a larger blob. Doing this for all overlapping blocks, the blocks of the meet are the subsets of U that are a union of certain blocks of π and simultaneously a union certain blocks of σ and are minimal in that respect.

The top of the lattice of partitions $\Pi (U)$ is the maximally distinguished *discrete partition* $\mathbf{1}_U = \{\{u_1\}, ..., \{u_n\}\}$ whose blocks are all the singletons of the elements of U and the bottom is the minimally distinguished *indiscrete partition* $\mathbf{0}_U = \{U\}$, nicknamed "The Blob,"[8] which blobs all the elements together into one indefinite "superposition."[9]

The *implication* $\sigma \Rightarrow \pi$ is the partition obtained from π where for each block $B \in \pi$, if there is a block $C \in \sigma$ such that $B \subseteq C$, then B is replaced by the discretized version of B, i.e., all singletons $\{u_i\}$ for $u_i \in B$, and otherwise, B remains the same in $\sigma \Rightarrow \pi$. Since the blocks of $\sigma \Rightarrow \pi$ are either the discretized versions of B, like $\mathbf{1}_B$, or are just B, like $\mathbf{0}_B$, then the implication $\sigma \Rightarrow \pi$ can be viewed as a 'characteristic function' for the inclusion of blocks of π in the blocks of σ with the values $\mathbf{1}_B$ or $\mathbf{0}_B$.

In the logic of subsets, the subset implication $S \Rightarrow T$ (or conditional $S \supset T$) for $S, T \subseteq U$ is the subset $S^c \cup T$ (where S^c is the complement $U - S$). The basic

---

[8] Like in the Hollywood movie of that name, "The Blob" absorbs everything it meets, i.e., $\mathbf{0}_U \wedge \pi = \mathbf{0}_U$.

[9] Many of the older texts (Birkhoff 1948) presented the "lattice of partitions" upside down, i.e., with the opposite partial order, so the join and meet as well as the top and bottom were interchanged.

property of the set implication is that it is a set that indicates the extent to which the partial order of inclusion holds for S in T so that when the implication is equal to the top of the Boolean algebra of subsets U, then actual inclusion holds, i.e.,

$$S \Rightarrow T = U \text{ if and only if (iff) } S \subseteq T.$$

The partition implication satisfies the corresponding property in the logic of partitions. It is the partition that indicates the extent to which σ is refined by π, i.e.,

$$\sigma \Rightarrow \pi = \mathbf{1}_U \text{ iff } \sigma \precsim \pi,$$

as is clear by viewing it as the 'characteristic function' for inclusion of the blocks of π in the blocks of σ.

There are several methods by which the usual Boolean operations on subsets can be systematically carried over to define the corresponding operations on partitions (Ellerman 2019). One of them is to just apply the Boolean set operations to ditsets to get the ditset for the corresponding partition operation. This works, for example, for the join since: dit $(\sigma \vee \pi) = $ dit $(\sigma) \cup$ dit (π). But the Boolean combination of ditsets is not necessarily a ditset so one needs the interior operation to obtain the largest ditset contained in an arbitrary subset of $U \times U$.

There is a natural closure operation on subsets $W \subseteq U \times U$, namely the reflexive-symmetric-transitive (RST) closure $\overline{W}$ of W which is the smallest indit set or equivalence relation containing W–which could also be taken as the intersection of all the equivalence relations containing W. Then an interior operation is obtained (as usual in topology) as the complement of the closure of the complement, i.e., int $(W) = \left(\overline{W^c}\right)^c$. When the Boolean combination of ditsets does not yield a ditset, then the interior yields the ditset of the corresponding operation on partitions. For instance, to define the partition meet, the subset meet operation on ditsets yields dit $(\sigma) \cap$ dit (π) which is not necessarily a ditset, but dit $(\sigma \wedge \pi) = $ int (dit $(\sigma) \cap$ dit (π))–which gives the same operation as defined previously. And for the partition implication:

$$\text{dit} (\sigma \Rightarrow \pi) := \text{int} \left(\text{dit} (\sigma)^c \cup \text{dit} (\pi)\right) = \text{int} (\text{indit} (\sigma) \cup \text{dit} (\pi))$$

which gives the same partition implication as previously defined (Ellerman 2010).

It might be noted that the RST-closure operation on subsets of $U \times U$ is not a topological closure operation since the union of two 'closed' subsets, i.e., two equivalence relations, is not necessarily an equivalence relation. If it were a topological closure operation, then the algebra of partitions (defined with the join, meet, and implication operations and the top and bottom) would be a Heyting algebra (which is always distributive)–but the lattice of partitions is not necessarily distributive.[10]

---

[10] In the early days of American mathematics, the logician and philosopher, Charles Saunders Peirce (1839–1914), claimed to have proved that all lattices are distributive but omitted the 'proof' as being too tedious (Peirce 1880). Europeans such as Dedekind and Schröder soon besieged him with counterexamples such as the simplest non-trivial partition lattice on a three-element set which is not distributive.

Of all the Boolean operations on partitions, the join operation is the main one we will need as it prefigures a projective measurement in quantum mechanics.

1.6 The Underlying Duality of Its and Dits

Underlying the duality between subsets (e.g., images of functions) and partitions (inverse-images of functions) is the duality between elements (its) of subsets and distinctions (dits) of a partitions, the 'its and dits' duality. The notion of a function is naturally defined using the dual notions of elements or its and distinctions or dits.

Given two sets X and Y, consider a binary relation $R \subseteq X \times Y$.

The relation R is said to *transmit elements* if for all $x \in X$, there is an ordered pair $(x, y) \in R$ for some $y \in Y$.

The relation R is said to *reflect elements* if for all $y \in Y$, there is an ordered pair $(x, y) \in R$ for some $x \in X$.

The relation R is said to *transmit distinctions* if for any $(x, y) \in R$ and $(x', y') \in R$, if $x \neq x'$, then $y \neq y'$.

The relation R is said to *reflect distinctions* if for any $(x, y) \in R$ and $(x', y') \in R$, if $y \neq y'$, then $x \neq x'$.

Then a binary relation R is a *function* if it is defined everywhere on X, i.e., transmits elements, and if it is single-valued, i.e., reflects distinctions. In this manner, the notion of duality in *Set* (the category of sets and functions), that provides the duality between subsets and partitions, can be traced back to elements and distinctions (Ellerman 2021c). It might also be noted that when R transmits elements and reflects distinctions so that it is a function $f : X \to Y$, then the two special types of functions, injective (one-to-one) and surjective (onto) are defined respectively as transmitting distinctions and reflecting elements. Each function $f : X \to Y$ has an associated subset, namely the image $f(X) \subseteq Y$, and an associated partition, namely the inverse-image or coimage $\left\{ f^{-1}(y) \right\}_{y \in f(X)}$ as shown in Fig. 1.1.

We have already encountered this duality in taking the quantitative measure of a partition as the number of dits just as the number of its is the quantitative measure of the size of a subset. In the Boolean lattice $\wp(U)$ of subsets, the partial order $S \subseteq T$ for $S, T \in \wp(U)$ is the inclusion of elements. In the lattice of partitions $\Pi(U)$, the refinement partial order $\sigma \precsim \pi$ is just the inclusion of dits, i.e., $\sigma \precsim \pi$ iff $\text{dit}(\sigma) \subseteq \text{dit}(\pi)$. The top U of the the subset lattice thus includes all possible

Fig. 1.1 Image of function is a subset of the codomain and coimage (or inverse-image) is a partition on the domain

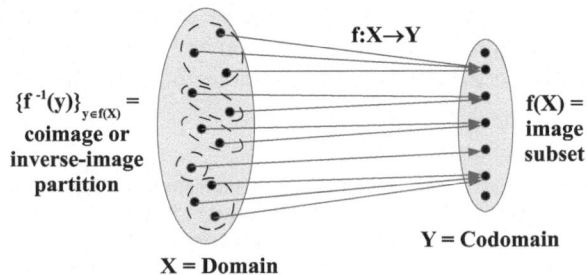

Table 1.2 Duality: $\frac{\text{Elements}}{\text{Subsets}} \approx \frac{\text{Distinctions}}{\text{Partitions}}$

Its & Dits	Lattice of subsets $\wp(U)$	Lattice of partitions $\Pi(U)$
Its or Dits	Elements of subsets	Distinctions of partitions
Partial order	Inclusion of elements $S \subseteq T$	Inclusion of dits dit $(\sigma) \subseteq$ dit (π)
Join	Union of elements $S \cup T$	Union of dits dit $(\pi \vee \sigma) =$ dit $(\pi) \cup$ dit (σ)
Top	Subset U with all elements	Partition $\mathbf{1}_U$ with all distinctions
Bottom	Subset $\emptyset$ with no elements	Partition $\mathbf{0}_U$ with no distinctions

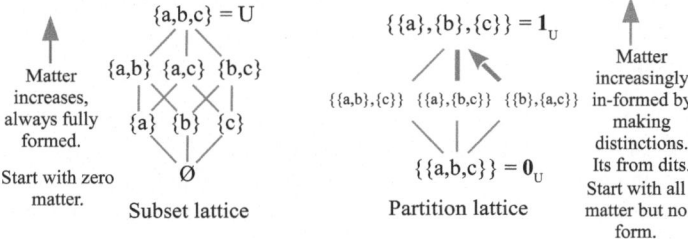

Fig. 1.2 The two dual notions of becoming illustrated by the two possible dual logical lattices

elements and the top $\mathbf{1}_U$ of the partition lattice includes all possible distinctions, i.e., dit $(\mathbf{1}_U) = U \times U - \Delta$ [where Δ is the diagonal of self-pairs (u_i, u_i)]. The bottom $\emptyset$ of the subset lattice has no elements and the bottom $\mathbf{0}_U$ of the partition lattice has no distinctions, i.e., dit $(\mathbf{0}_U) = \emptyset$. The elements of the join in the subset lattice is the union of the elements, and the dits of the join in the partition lattice is the union of the dits. In mathematics itself, the dual lattices, the lattices of subsets and of partitions, are equally fundamental (Ellerman 2021b). This duality between elements and distinctions is illustrated in Table 1.2.

Following Heisenberg, we might express this duality by going back to the ancient Greek metaphysical notions of matter (or substance) and form (Ainsworth 2020). At this simple level, one can still discern two modes of becoming or creation corresponding to the subset version (always fully definite) and partition (increasing definiteness) version. These two stories can be represented by moving from the bottom up the two logical lattices illustrated in Fig. 1.2 where the universe consists three states $U = \{a, b, c\}$.

At the bottom of the subset lattice is the empty set which represents no matter (no elements or its). Then the process of becoming or creation takes place by the addition of new fully-formed elements or its in bigger and bigger subsets until reaching the universe set at the top.

At the bottom of the partition lattice is the Blob where all the matter exists but in a completely indefinite form (no dits), and then "becoming" takes place as more and more distinctions are made moving up the lattice until reaching the top partition where all the distinctions have been made. As a metaphor for the Big Bang, in the beginning there was all the matter or energy with "perfect symmetry" (Pagels 1985), and then distinctions were introduced in the form of symmetry-breaking.

In general, this is the partitional notion of becoming, i.e., moving from an indefinite state, e.g., $\{\{b\}, \{a, c\}\}$, to a more definite state, e.g., $\{\{a\}, \{b\}, \{c\}\}$ (see the arrow in Fig. 1.2), by the in-forming of distinctions (i.e., "more definite its from dits"), e.g., the distinctions of $\{\{a\}, \{b, c\}\}$, that is key to understanding the quantum level notion of measurement. Put the other way around, more definite states *emerge* from less definite states by the making of distinctions. This in-forming of indefinite states to make them more definite is the skeletal math behind Heisenberg's sympathetic interpretation of Aristotle's treatment of matter and form where matter is "a kind of indefinite corporeal substratum, embodying the possibility of passing over into actuality by means of the form" (Heisenberg 1962, p. 148). The "form" is in-form-ation-as-distinctions as in the logical information theory sketched above.

Some metaphors might help.

- Classical change, the continuous change from definite to definite, is like flipping through a police mug book of fully definite faces.
- The quantum notion of becoming is like the police artist's sketch pad whose state changes from the indefinite outline of a face to becoming more and more definite until it is a realistic face.
- Another metaphor is a painter's canvas that is a superposition of all colors (i.e., white) and then slowly becomes a painting as the white spaces take on a definite coloring.
- Yet another metaphor is 3D printing where higher levels of increasing definiteness correspond to the upper levels of 3D printing as the printed object slowly takes on a definite shape.

Refining partitions is the mathematics for this transition from indefinite to more definite. The logical operation on partitions that results in a more refined partition is the *join* operation (or, equivalently, the intersection of equivalence relations) and, as we will see, the join operation is the skeletonized model (the set level version) of the projective measurement operation in QM.

1.7 Ontology: Quantum Reality as Indefinite World

It cannot be expected to arrive at a *classical* intuitive picture of quantum level reality. But one can develop some piecemeal imagery or conceptual crutches to help provide a better picture. For instance, how should one imagine a quantum superposition? The most misleading imagery in QM is the *classical* interpretation of superposition (Fig. 1.3) as the addition of two waves to get another wave, e.g., in the classroom ripple-tank model of the double-slit experiment. And there is superposition in classical electromagnetic theory.

The "Wave Interpretation" arises since the complex numbers are the natural mathematics to describe waves so the misleading wave imagery is always there. The wave mathematics is not wrong but it should not be interpreted as representing physical waves. And Dirac, among others, warned against the physical interpretation of superposition of waves.

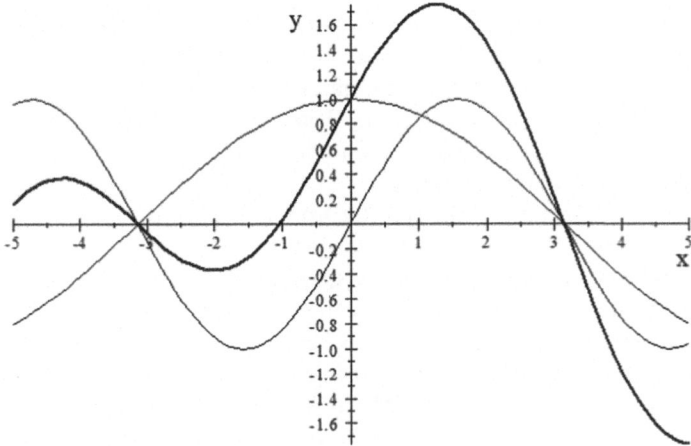

Fig. 1.3 Imagery for *classical* superposition

> Such analogies have led to the name 'Wave Mechanics' being sometimes given to quantum mechanics. It is important to remember, however, that the superposition that occurs in quantum mechanics is of an essentially different nature from any occurring in the classical theory, as is shown by the fact that the quantum superposition principle demands indeterminacy in the results of observations in order to be capable of a sensible physical interpretation. The analogies are thus liable to be misleading. (Dirac 1958, p. 14)

The complex numbers are needed in the mathematics of QM (among other reasons Aaronson 2013, Chap. 9) since the complex numbers are the algebraically-complete extension of the reals so the real-valued observables will have a full set of eigenvectors, (Weinberg 2015, p. 67, fn. 7) *not* because the wave function is an ontic wave. As R. I. G. Hughes pointed out, it is just a mathematical fact that the addition (superposition) of two vectors (in a vector space over $\mathbb{C}$) can *always* be expressed in terms of the interference of waves.

> The wave formalism offers a convenient mathematical representation of this latency, for not only can the mathematics of wave effects, like interference and diffraction, be expressed in terms of the addition of vectors (that is, their linear superposition; see (Feynman 1963, Chap. 29.5), but the converse, also holds. (Hughes 1989, p. 303)

The logical sequence is that the indefiniteness of superposition is mathematically represented by the addition of scalar-multiples of complex-valued eigenvectors in Hilbert space. Then those sums of vectors over the complex numbers can always be interpreted in terms of interference and diffraction of mathematical (not physical) waves as noted by R. I. G. Hughes and Richard Feynman. Quantization shows how the physics is represented or encoded in those mathematical waves, e.g., the De Broglie and Einstein formulas relating physical quantities like momentum and energy to wave variables like wavelength and frequency. That is how the indefiniteness of superposition ultimately and rather misleadingly is represented in the wave mathematics of QM.

One point is that the mathematical waves represent probability amplitudes, not 'matter waves.' "[I]t must be emphasized that the wave function that satisfies the

[Schrödinger] equation is not like a real wave in space; one cannot picture any kind of reality to this wave as one does for a sound wave" (Feynman 2010, Sect. 3.7). In this regard, the classroom demonstration of the double-slit experiment using the water ripple tank is seriously misleading. Water waves are matter waves; the wave function is not. And the particle does not go through both slits at once–as the ripple tank water wave is pictured as doing.

How does quantum superposition differ from the superposition of water or electromagnetic waves? What is the different interpretation of superposition? When definite- or eigen-states are superposed in QM, the result is a state that is indefinite where the definite states differ. In the quantum literature, this notion of a superposition is described variously as blurry, unsharp, smudged, blunt, fuzzy, blob-like, dispersed, smeared out, indeterminate, spread-out, cohered, or indefinite. Referring to a quantum particle as a "quanton," Mario Bunge makes this point about quantum superposition in comparison with the superposition of waves in classical physics.

> Another surprising peculiarity of quantons is that they are blurry or fuzzy rather than neat or sharp. Whereas in classical physics all properties are sharp, in quantum physics only a few are: most are blunt or smudged. ... The reason for this fuzziness is that ordinarily an isolated quanton is in a "coherent" state, that is, the combination or superposition (weighted sum) of two or more basic states (or eigenfunctions). The superposition or "entanglement" of states is a hallmark of quantum mechanics. (Bunge 2010, pp. 49–50)

The superposition in classical wave theories does not have that indefiniteness interpretation.

The density matrix formulation of QM brings this indefiniteness interpretation (e.g., the non-zero off-diagonal elements of a density matrix representing coherences) more into the foreground. The density matrix formulation tells us that the quantum world is Indefinite World. Einstein famously said that "the Lord is subtle, but not malicious." But the century-long formulation of the so-called "wave mechanics" using the probability-computational device of the wave function reveals that Einstein's Lord, while not being outright malicious, is at least a trickster.

The important point here is that, as will be shown, the math of QM is the Hilbert space version of the math of partitions–which is the math of indefiniteness and definiteness. The quantum interpretation of superposition is the addition of definite eigenstates of a particle to get a new state *indefinite between the more definite superposed eigenstates*–not the 'double-exposure' image suggested by the wave interpretation.[11] The mental shift from superposition-as-wave-addition (like in a ripple tank) to superposition-as-indefiniteness is the first important reconception necessary to understand the partition analysis.

In Fig. 1.4, the superposition of the two definite isosceles triangles is the indefinite triangle which is indefinite on where the two definite triangles are different (the labeling of the equal sides and equal angles) and is definite on where the two triangles are the same (the side A, the vertex a, and the aA-axis).

---

[11] This indefiniteness interpretation is now more common in quantum information and computation theory (Jaeger 2007; Nielsen and Chuang 2000) where a qubit $\alpha\,|0\rangle + \beta\,|1\rangle$ is not interpreted as

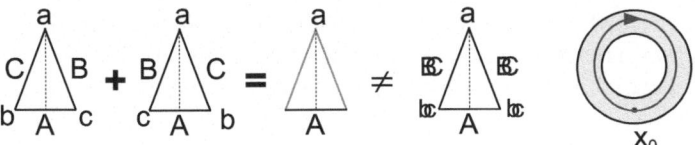

Fig. 1.4 Imagery for *Quantum* Superposition

David Z. Albert has pointed out the little-noticed fact that a definiteness that is common to all the components in a superposition will be definite in the superposition state.

> It follows from the linearity of the operators which represent observables of quantum mechanical systems that any measurable physical property which happens to be shared by all of the individual mathematical terms of some particular superposition (written down in any particular basis) will necessarily also be shared by the full superposition, considered as a single quantum-mechanical state, as well. (Albert 1997, p. 234)

This also confirms that the notion of a superposition and the notion of abstraction (i.e., abstracting to the properties common to the elements in a set) are 'essentially' that same concept viewed from flip-side viewpoints. In QM, the emphasis is on the indefiniteness of the properties that are not the same in the elements of the set (the glass is half-empty) versus the emphasis in abstraction on the definiteness of the properties that are common to the elements in the set (the glass is half-full) (Ellerman 2021a). In Fig. 1.4, the superposition triangle could also be viewed as the abstraction from the two isosceles triangles to represent their common features (the vertex *a* and the side *A*).

In the older homotopy theory, the abstraction of a "homotopy type" was represented by a block in a partition, i.e., by an equivalence class. "Homotopy types are equivalence classes of spaces" under the equivalence relation "of deformation or homotopy" (Baues 1995, p. 4). In the modern treatment (Univalent Foundations Program 2013), the equivalence class of, say, unit-interval-coordinatized paths going once around the ring clockwise is abstracted to an object having those common characteristics (Fig. 1.4) but is not itself one of the coordinatized paths in the equivalence class.[12] This connection between *superposition as indefiniteness-on-differences* and *abstraction as definiteness-on-commonalities* is unavailable to those thinking of the notion of superposition as classical wave-addition.[13]

---

being simultaneously $|0\rangle$ and $|1\rangle$, like a 'wave' simultaneously at two slits, but as a state indefinite between $|0\rangle$ and $|1\rangle$.

[12] In Frege's example of a set of parallel directed line segments oriented in the same way, the abstraction "direction" is definite on what is common between the lines and indefinite on how they differ (Ellerman 2021a).

[13] The connection between superposition and the mental process of abstraction has been developed with great energy and imagination by Diederik Aerts and colleagues (Aerts 2010; Aerts and Sassoli de Bianchi 2017; Sassoli de Bianchi 2021).

What is needed is an equivalent alternative to the mathematically correct but misleading wave function notion of quantum state that fits precisely with the partition approach. There is such an alternative mathematical formulation in QM, namely density matrices.

References

Aaronson S (2013) Quantum computing since Democritus. Cambridge University Press, New York

Aerts D (2010) A potentiality and conceptuality interpretation of quantum physics. Philosophica 83:15–52

Aerts D, Sassoli de Bianchi M (2017) Quantum measurements as weighted symmetry breaking processes: the hidden measurement perspective. Int J Quantum Found 3:1–16

Ainsworth T (2020) Form versus Matter. In: Zalta EN (ed) The Stanford encyclopedia of philosophy (Summer 2020 Edition)

Albert DZ (1997) What superpositions feel like. In: Earman J, Norton JD (eds) The cosmos of science: essays of exploration. University of Pittsburgh Press, Pittsburgh, pp 224–242

Ariew R (ed) (2000) G W Leibniz and Samuel Clarke: correspondence. Hackett, Indianapolis

Auletta G (2019) The quantum mechanics conundrum: interpretations and foundations. Springer Nature, Cham, Switzerland

Baues H-J (1995) Homotopy Types. In: James IM (ed) Handbook of algebraic topology. Elsevier Science, Amsterdam, pp 1–72

Birkhoff G (1948) Lattice theory. American Mathematical Society, New York

Boole G (1854) An investigation of the laws of thought on which are founded the mathematical theories of logic and probabilities. Macmillan and Co, Cambridge

Britz T, Mainetti M, Pezzoli L (2001) Some operations on the family of equivalence relations. In: Crapo H, Senato D (eds) Algebraic combinatorics and computer science: a tribute to Gian-Carlo Rota. Springer, Milano, pp 445–59

Bunge M (2010) Matter and mind: a philosophical inquiry. Spring Publications, Dordrecht

Cushing JT (1988) Foundational problems in and methodological lessons from quantum field theory. In: Brown HR, Harre R (eds) Philosophical foundations of quantum field theory. Clarendon Press, Oxford, pp 25–39

Dirac PAM (1958) The principles of quantum mechanics, 4th edn. Clarendon Press, Oxford

Dubreil P, Dubreil-Jacotin M-L (1939) Théorie algébrique des relations d'équivalence. J de Mathématique 18:63–95

Ellerman D (2010) The logic of partitions: introduction to the dual of the logic of subsets. Rev Symb Logic 3:287–350. https://doi.org/10.1017/S1755020310000018

Ellerman D (2014) An introduction to partition logic. Logic J IGPL 22:94–125. https://doi.org/10.1093/jigpal/jzt036

Ellerman D (2019) A graph-theoretic method to define any Boolean operation on partitions. Art Discret Appl Math 2:1–9. https://doi.org/10.26493/2590-9770.1259.9d5

Ellerman D (2021a) On abstraction in mathematics and indefiniteness in quantum mechanics. J Philosop Logic 50:813–835. https://doi.org/10.1007/s10992-020-09586-1

Ellerman D (2021b) New foundations for information theory: logical entropy and Shannon entropy. Springer Nature, Cham, Switzerland

Ellerman D (2021c) The logical theory of canonical maps: the elements and distinctions analysis of the morphisms, duality, canonicity, and universal constructions in set. https://arxiv.org/abs/210408583

Ellerman D (2023) The logic of partitions: with two major applications. College Publications, London

Feynman RP, Leighton RB, Sands M (1963) The Feynman lectures on physics: mainly mechanics, radiation, and heat, vol I. Addison-Wesley, Reading, MA

Feynman RP, Leighton RB, Sands M (2010) The Feynman lectures on physics: quantum mechanics vol. III (New Millennium Ed). Addison-Wesley, Reading, MA

Finberg D, Mainetti M, Rota G-C (1996) The logic of commuting equivalence relations. In: Ursini A, Agliano P (eds) Logic and algebra. Marcel Dekker, New York, pp 69–96

Heisenberg W (1962) Physics and philosophy: the revolution in modern science. Harper Torchbooks, New York

Hughes RIG (1989) The structure and interpretation of quantum mechanics. Harvard University Press, Cambridge

Jaeger G (2007) Quantum information: an overview. Springer Science+Business Media, New York

Jaeger G (2019) Information and the reconstruction of quantum physics. Annalen der Physik 531:1800097. https://doi.org/10.1002/andp.201800097

Kant I (1998) Critique of pure reason. Cambridge University Press, Cambridge, UK

Kung JPS, Rota G-C, Yan CH (2009) Combinatorics: the Rota way. Cambridge University Press, New York

Lawvere FW, Rosebrugh R (2003) Sets for mathematics. Cambridge University Press, Cambridge, MA

Leibniz GW (1996) New essays on human understanding. Cambridge University Press, Cambridge, UK

Lewis PJ (2016) Quantum ontology: a guide to the metaphysics of quantum mechanics. Oxford University Press, New York

Nielsen M, Chuang I (2000) Quantum computation and quantum information. Cambridge University Press, Cambridge

Pagels H (1985) Perfect symmetry: the search for the beginning of time. Simon and Schuster, New York

Peirce CS (1880) On the algebra of logic. Am J Math 13:15–57

Rota G-C (1998) Combinatorial theory: the Guidi notes. MIT Copy Services, Cambridge, MA

Rota G-C (2001) Twelve problems in probability no one likes to bring up. In: Crapo H, Senato D (eds) Algebraic combinatorics and computer science: a tribute to Gian-Carlo Rota. Springer, Milano, pp 57–93

Sassoli de Bianchi M (2021) A non-spatial reality. Found Sci 26:143–170. https://doi.org/10.1007/s10699-020-09719-4

Shimony A (1988) The reality of the quantum world. Sci Am 258:46–53

Univalent Foundations Program (2013) Homotopy type theory: univalent foundations of mathematics. Princeton, Institute for Advanced Studies

Valentini A (2011) Quantum Interview. In: Schlosshauer M (ed) Elegance and enigma: the quantum interviews. Springer, Heidelberg

Weinberg S (1994) Dreams of a final theory. Vintage Books, New York

Weinberg S (2015) Lectures on quantum mechanics, 2nd edn. Cambridge University Press, Cambridge, UK

Whitman PM (1946) Lattices, equivalence relations, and subgroups. Bull Am Math Soc 52:507–522

Wigner EP (1967) The unreasonable effectiveness of mathematics in the natural sciences. Symmetries and reflections. Indiana University Press, Bloomington, IN, pp 222–237

Chapter 2
Partitions: The Logical Concept to Describe Indefiniteness and Definiteness

Any blurring in classical physics is statistical and can be accounted for, in principle, because it is associated with sharp substructures. In contradistinction, in the quantum world we are forced to accept blurring as a fundamental ontic feature that, in principle, cannot be accounted for by averaging over sharp substructures. Quantum objects with blurred features therefore really exist.

Fritz Rohrlich (1986, p. 380)

Abstract The basic non-classical notion in quantum mechanics (QM) is the notion of superposition; entanglement is a particularly vexing special case. This chapter focuses on slowly developing the notion of a superposition state starting at the logical level of superposition applied to the notion of a set. To mathematically represent a superposition set, one must go beyond the one-dimensional notion of a 0, 1-vector representing which elements of the universe set are in the subset. A two-dimensional matrix will do the job with the 0, 1-vector along the diagonal and the off-diagonal elements of 1 representing which elements are superposed in the sense of being in the same equivalence class of an equivalence relation (which is just another way to look at a partition). Those 0, 1-matrices are the incidence matrices of equivalence relations. When such logical representations of equivalence relations are normalized by dividing through by the trace (sum of the diagonal elements), one arrives at a density matrix—which is the partitional way to represent a quantum state. The three main distinctive elements of QM math are the notions of a quantum state, a quantum observable, and a (projective) measurement applying an observable to a state. This chapter gives the partition-based treatment of quantum states via density matrices.

Keywords Superposition · Logical entropy · Quantum states · Density matrices

© The Author(s), under exclusive license to Springer Nature Switzerland AG 2024
D. Ellerman, *Partitions, Objective Indefiniteness, and Quantum Reality*,
SpringerBriefs in Philosophy, https://doi.org/10.1007/978-3-031-61786-7_2

2.1 Superposition Applied to Subsets

We assume an ontology of quantum particles (or "quantons" (Levy-Leblond and Balibar 1990; Bunge 2010)) but they are not the particles of classical (or Bohmian) mechanics since they can be in indefinite superposition states. The so-called "particle/wave complementarity" is the "complementarity" between a particle being in a definite eigenstate versus a particle in an indefinite superposition state represented by the state vector or by a density matrix (see below).

At the simplest logical level, the pure and mixed quantum states for a single particle can be illustrated by partitions as in Fig. 2.1. Suppose there are four possible eigenstates (or definite states) for a particle represented by a, b, c, and d. The pure state of all those eigenstates superposed is represented by the indiscrete partition $\{\{a, b, c, d\}\}$ (written in shorthand as $\{abcd\}$ which eliminates the inner-most curly brackets in favor of juxtaposition) and then distinctions are made (i.e., 'measurements' are made) to get the other partition representations of the mixed states of 'orthogonal' (disjoint) superpositions such as $\{\{a, c\}, \{b, d\}\}$ (or in shorthand $\{ac, bd\}$). The lines between partitions are the refinement relations. Figure 2.1 is a Hasse diagram for the lattice of partition on U where the connecting lines between partitions are refinement relations and there is no partition intermediate between the partitions connected by a line. And there is no notion of continuous evolution from a block of one partition to a smaller block of a more refined partition; it is a discontinuous probabilistic jump.[1]

A quantum superposition such as $\alpha|a\rangle + \gamma|c\rangle$ is *skeletonized* by throwing away the non-zero scalars, ket symbols, and the addition to just obtain the subset $\{a, c\}$. If we define the *support* of a vector expressed in a certain basis as the set of basis vectors with non-zero coefficients, then skeletonizing just takes a vector to its support. In that manner one obtains a skeletal picture of some classical and quantum states as in Fig. 2.1 for the universe set of states $U = \{a, b, c, d\}$ of a particle. The mixed states all come from making distinctions in the pure state at the bottom.

In set theory, this is the jump from a set (e.g., an equivalence class) to an element of the set (as in a maximal measurement) given by a "choice function"

Fig. 2.1 Lattice of partitions as a skeletal view of some classical and quantum states

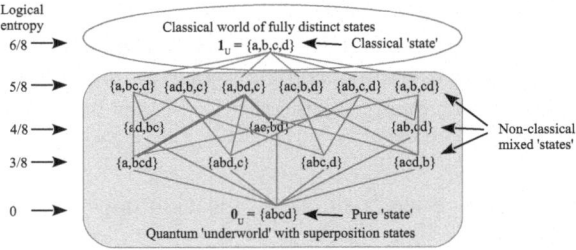

Table 2.1 Set and Quantum versions of non-determinate and determinate results

Choice function	Quantum measurement (maximal)
$f(\{a, b, ..., c\}) = b$	$\alpha\lvert a\rangle + \beta\lvert b\rangle + \cdots + \gamma\lvert c\rangle \overset{\text{indeterminate}}{\rightsquigarrow} \lvert b\rangle$
$f(\{b\}) = b$	$\lvert b\rangle \overset{\text{determinate}}{\rightsquigarrow} \lvert b\rangle$

(Halmos 1974, p. 60) which is non-deterministic for non-singleton sets.[2] A singleton set represents an eigenstate and the measurement of an eigenstate of the observable being measured has no indeterminism; the outcome is the same eigenstate with probability one as indicated in Table 2.1.

The choice function shows how even the indeterminism in the QM case is naturally represented at the skeletal level of sets.

In terms of sets (as in Fig. 2.1), the superposition of the eigenstates $\{a\}$ and $\{c\}$ is the state $\{a, c\}$, which is the state indefinite between $\{a\}$ and $\{c\}$; not definitely $\{a\}$ and not definitely $\{c\}$, but definitely not $\{b\}$ or $\{d\}$. A distinction between $\{a\}$ and $\{c\}$, e.g., in the join $\{ac, bd\} \vee \{a, bcd\} = \{a, bd, c\}$, would reduce the superposition $\{a, c\}$ (or $\{ac\}$ in shorthand) to a mixture of the eigenstates $\{a\}$ and $\{c\}$, e.g., in the mixed state $\{a, bd, c\}$.

We will see later how to go (via the Yoga of Linearization) from the skeletal set level notions, such as a non-singleton block or subset in a partition to the corresponding vector (Hilbert) space notion of a superposition state.

One of the important 'takeaways' from the partitional analysis is that there are levels of indefiniteness below the classical level represented by the discrete partition. And unitary evolution of a quantum state does not have to rise to the classical level (where our human-scale detection devices operate) but can proceed at a lower level of definiteness. In the double-slit experiment, the evolution of the superposition state, |going through slit 1⟩ + |going through slit 2⟩, in the case of no detection at the slits, takes place at a non-classical level of indefiniteness; the results are only detected when the locations of the particles hitting the detection wall are measured at the classical level.

QM has the partitional notion of becoming or change–from indefinite to less indefinite. Since the detection of a particle's location can only take place at the classical level while unitary evolution can take place at any level of indefiniteness, the classical notion of a particle's "trajectory" does not apply. As David Hawkins put it, the picture at the quantum level "is no longer that of a steady deterministic flow..., but that of states and transitions, 'flights and perchings,' in which the perchings are more stable and flights more abrupt than classical ideas would have allowed." (Hawkins 1964, p. 198).

There is a strong ontological hint in the rather standard view that a superposition state (in the measurement basis) does not have a definite value prior to the mea-

---

[2] This indeterminacy in a choice function is why the Axiom of Choice (which states that every nonempty set has a choice function) has to be added as an *independent* axiom in axiomatic set theory. In general, a choice function cannot be defined by other set theoretic operations.

surement. The Heisenberg Indeterminacy Principle shows that conjugate variables cannot both have definite values. Some quantum theorists, starting with Heisenberg, have taken the hint so the idea of different levels of indefiniteness below the classical level is not at all new–but it was previously expressed in the language of potentialities (or latencies) versus actualities (Heisenberg 1962; Margenau 1954; Herbert 1985, Quantum reality #8 Hughes 1989, Sect. 10.2; Shimony 1999; Fleming 1992; Karakostas 2007; Kastner et al. 2018; kastner 2013, 2015; Chiatti 2022; Kožnjak 2007; deRonde 2018; Jaeger 2014, 2017; Strumia 2021; Del Santo and Gisin 2023).

> Heisenberg (1962, p. 53)… used the term "potentiality" to characterize a property which is objectively indefinite, whose value when actualized is a matter of objective chance, and which is assigned a definite probability by an algorithm presupposing a definite mathematical structure of states and properties. Potentiality is a modality that is somehow intermediate between actuality and mere logical possibility. That properties can have this modality, and that states of physical systems are characterized partially by the potentialities they determine and not just by the catalogue of properties to which they assign definite values, are profound discoveries about the world, rather than about human knowledge. (Shimony 1999, p. 6)

The quantum theorists who use the potentiality-to-actuality language are referring to the actualization of a more definite state starting with an indefinite superposition state. Hence a more precise usage would be the indefinite-to-definite transition between two states of reality. Shimony makes this point when discussing Heisenberg's reference to Aristotle's notion of "potentia."

> The historical reference should perhaps be dismissed, since quantum mechanical potentiality is completely devoid of teleological significance, which is central to Aristotle's conception. What it has in common with Aristotle's conception is the indefinite character of certain properties of the system. (Shimony 1993, pp. 313–4)

Henry Margenau and R. I. G. Hughes favor the term "latency" over "potentiality" but Margenau mentions that the measurement of observables "forces them out of indiscriminacy or latency" (Margenau 1954, p. 10)–which indicates that Margenau also interprets "latency" in terms of indeterminacy or indefiniteness. And Ruth Kastner considers the indeterminacy of values as a characteristic of the real potentia (Kastner 2013, p. 3).[3] "Although the quantum possibilities represent properties that might be observed, they are not determinate (that is, they are not well defined)." (Kastner 2015, p. 27). The analysis in terms of "potentia" (or "latencies") versus actualities should be reformulated in terms of objective indefiniteness and definiteness. The relevant logic is not that of modalities (possible versus actual) but the logic of partitions, the mathematical concept to describe indistinctions and distinctions. In short, it is not real potentialities versus real actualities but real (objective) indefiniteness versus real definiteness. Rohrlich (1986, 1987) directly adopted the indefiniteness or ontic blurred interpretation of quantum reality and only interpreted the talk of potentialities as a manner of speaking (like saying that a thrown die, while in the air, has the

---

[3] Kastner likens reality to an iceberg with the part above the water is the reality of space-time while the rest or the iceberg stands for the quantum realm of potentialities. In short, the iceberg (Kastner 2013, p. 3) is akin to the lattice of partitions where the discrete partition at the top "above water" represents classical fully definite reality. See below for the partition logical Principle of Identity of Indistinguishables.

Table 2.2 Corresponding partition and quantum concepts

Partition concept	Corresponding quantum concept
Non-singleton block, e.g., $\{a, b, c\}$	Superposition pure state
Indiscrete partition $\mathbf{0}_U$	Largest pure state
Singleton block, e.g., $\{d\}$	Classical state (no superposition)
Discrete partition $\mathbf{1}_U$	Complete mixture of classical states

potentiality of coming up with a five), not as an ontic category. Peter Mittelstaedt noted that "the quantum mechanical unsharpness of measurable properties must be considered as an objective indeterminateness" (Mittelstaedt 1998, p. 178) and Paul Feyerabend remarked that the "inherent indefiniteness is a universal and objective property of matter." (Feyerabend 1983, p. 202).

The potentiality-actuality interpretation of quantum reality also does not provide the criterion for actualization. On the indefiniteness interpretation, the jump from an indefinite state of a more definite or fully definite state is accomplished by the making of distinctions (more-definite its from dits). Starting with a pure state, the making of distinctions leads to a probabilistic mixed state, i.e., Lüders mixture operation, and then the final outcome state, i.e., Lüders rule, is given by those probabilities. Moreover, as we will later see, the indefiniteness-definiteness (as opposed to the potentiality-actuality) formulation meshes perfectly with the notion of information-as-distinctions measured by logical entropy and with the Feynman rules based on indistinguishability and distinguishability.[4] Since ontological indefiniteness is already widely recognized (e.g., the Indeterminacy Principle), Occam's Razor should be applied to the language of real but non-actual potentialities.

The brain-twisting reconceptualization needed to understand quantum-level reality is to discard the classical view of only fully definite reality in favor an underlying reality of objective indefiniteness where processes take two forms: (1) the state-reduction jump from an indefinite state to a more definite state due to the making of distinctions ("more-definite its from dits"), and (2) change at the same level of indefiniteness (represented in QM math by unitary evolution). In the classical view of fully definite reality, there is only one type of change, definite to definite, given by classical mechanics.

It is important to emphasize that the partition math approach does not create a new interpretation of QM but adds corroboration and detail to the (suitably reformulated in terms of indefiniteness) views of Heisenberg, Shimony, Jaeger, and the others who have already posited a quantum world characterized by objective indefiniteness–the Indefinite World. The partitional concepts at the set level and the corresponding quantum concepts are paired in Table 2.2:

---

[4] Indeed, the whole translation dictionary developed in mathematical terms between partition math and QM math is based on the distinctions versus indistinctions concepts and is unavailable on the potentialities interpretation.

Is there some equivalent mathematical tool, instead of the wave-function (or state vector) that directly represents indefiniteness? Yes, it is the density matrix, but we will approach that concept gradually by first considering how one might define the notion of a 'superposition event' in 'almost-classical probability theory' in contrast to the usual notion of a discrete event of fully distinct elements in classical probability theory.

One way to better understand the indefiniteness interpretation of superposition in QM is to see how a notion of superposition could be formally developed at the simpler level of sets, and how this intuitive idea of a *superposition subset* can be mathematically described. If $U = \{a, b, c\}$, then the discrete partition on U could be visualized as an equilateral triangle ${}_b\triangle_c^a$ with all the vertices distinguished. But a partition $\{\{a, c\}, \{b\}\}$ has a superposition subset symbolized as $\Sigma\{a, c\}$ and perhaps visualized as:

$$ {}_b\triangle_?^? = {}_b\triangle_c^a + {}_b\triangle_a^c $$

which is indefinite on differences and definite on commonalities.

How can the classical notion of a subset $\{a, c\}$, consisting the distinct states a and c, be mathematically distinguished from the notion of the superposition subset $\Sigma\{a, c\}$, at the level of sets? Partition math already recognizes that difference as between the classical $\{\{a\}, \{c\}\}$ (fully distinguished states with no superposition) and the superposition $\{a, c\}$. In the context of a partition, we always interpret a block with multiple elements as a superposition subset. But we need to prefigure the *two*-dimensional notion of a matrix such as the density matrices developed later.

We take the universe U to be the following set of polygon shapes as in Fig. 2.2.

The figures have two properties, the number of sides and being solid s or hollow h, so the universe could be characterized as $U = \{u_1, u_2, u_3, u_4\} = \{3\,h, 3\,s, 4\,s, 5\,s\}$. The subset S of solid figures could be represented as a column vector: $|S\rangle = [0, 1, 1, 1]^t$ (where $()^t$ signifies the transpose operation) with the given ordering $[3\,h, 3\,s, 4\,s, 5\,s]^t$. By moving up one dimension, we can find the appropriate mathematical representation for both the classical subset S *and* the superposition subset ΣS.

For a universe of n elements, $U = \{u_1, ..., u_n\}$, the *incidence matrix* $\text{In}(R)$ of a binary relation $R \subseteq U \times U$ is the $n \times n$ matrix with $\text{In}(R)_{jk} = 1$ if $(u_j, u_k) \in R$ and 0 otherwise. The incidence matrix of a partition expressed as the corresponding equivalence relation is, when normalized, a simple type of density matrix.

We have two cases for a subset $S \subseteq U$.

1. The ordinary notion of the subset S (i.e., set of solid figures) is represented by the diagonal incidence matrix $\text{In}(\Delta S)$ that lays the column vector $|S\rangle$ along the

Fig. 2.2 U as a set of equilateral polygons

$$ U = \{\triangle, \blacktriangle, \blacksquare, \pentagon\} $$

$$
\text{diagonal: In}(\Delta S) = \begin{bmatrix} 0 & 0 & 0 & 0 \\ 0 & 1 & 0 & 0 \\ 0 & 0 & 1 & 0 \\ 0 & 0 & 0 & 1 \end{bmatrix} = \text{representation of set } S \text{ of distinct } S\text{-entities.}
$$

In(ΔS) is the incidence matrix of the diagonal relation $\Delta S = \{(u_i, u_i) : u_i \in S\} \subseteq U \times U$.

2. The representation of the superposition subset ΣS is the incidence matrix In$(S \times S)$ for the relation $S \times S$. This $n \times n$ incidence matrix can also be obtained as the outer product of the $n \times 1$ column vector $|S\rangle$ times the $1 \times n$ row vector $(|S\rangle)^t$:

$$
\text{In}(S \times S) = |S\rangle(|S\rangle)^t = \begin{bmatrix} 0 & 0 & 0 & 0 \\ 0 & 1 & 1 & 1 \\ 0 & 1 & 1 & 1 \\ 0 & 1 & 1 & 1 \end{bmatrix} = \text{representation of indefinite superposition}
$$

state $\Sigma S = \{3\,s, 4\,s, 5\,s\}$.

Intuitively, there is no distinction between a singleton subset such as $S^c = \{3h\}$ and the corresponding superposition subset since a singleton has no multiple elements to blob together in a superposition, and accordingly the two incidence matrices are the same for singletons.

The difference between the two representations of the subset S is in the off-diagonal entries. Think of the non-zero off-diagonal In$(S \times S)_{jk} = 1$'s for $j \neq k$ as equating, cohering, blurring out, 'blobbing' together, or making indefinite the differences (e.g., the number of sides) between u_j and u_k which have the common property of 'being a solid figure' so a 1 in the incidence matrix represents an indistinction of the equivalence class $\overset{S}{\sim}$ of being solid:

$$
\text{In}(S \times S) = \begin{bmatrix} 0 & 0 & 0 & 0 \\ 0 & 1 & 1 & 1 \\ 0 & 1 & 1 & 1 \\ 0 & 1 & 1 & 1 \end{bmatrix} \text{ says } \begin{bmatrix} 0 & 0 & 0 & 0 \\ 0 & 3s \overset{S}{\sim} 3s & 3s \overset{S}{\sim} 4s & 3s \overset{S}{\sim} 5s \\ 0 & 4s \overset{S}{\sim} 3s & 4s \overset{S}{\sim} 4s & 4s \overset{S}{\sim} 5s \\ 0 & 5s \overset{S}{\sim} 3s & 5s \overset{S}{\sim} 4s & 5s \overset{S}{\sim} 5s \end{bmatrix}.
$$

Intuitively, the differences in the number of sides of the solid figures have been blurred out or rendered indefinite in the representation of the superposition. Our thesis is that superposition means indefiniteness, and this comes out even when dealing just with incidence matrices–prior to the parallel development with density matrices. That is, when there are two or more elements in the subset S represented by the column vector $|S\rangle$, then there are the non-zero off-diagonal elements representing indefiniteness in the incidence matrix.

At the full quantum level in QM/$\mathbb{C}$, the superposition of the three 'solid' states would have complex scalar coefficients as in: $\alpha|3\,s\rangle + \beta|4\,s\rangle + \gamma|5\,s\rangle$ (where $\alpha\alpha^* + \beta\beta^* + \gamma\gamma^* = 1$ where $\alpha^* = x - yi$ is the complex conjugate of a complex number $\alpha = x + yi$). But, at the skeletal set level, the scalars are stripped away to leave the notion of the superposition subset represented by In$(S \times S)$–which may be juxtaposed to the ordinary subset represented by In(ΔS) which at the quantum level would be a mixed state $|3s\rangle$, $|4s\rangle$, and $|5s\rangle$ with the respective probabilities

$\Pr(3s)$, $\Pr(4s)$, and $\Pr(5\,s)$ that sum to 1. These two different incidence matrices, $\text{In}(S \times S)$ and $\text{In}(\Delta S)$, prefigure two different density matrices (to be explained later) at the quantum level:

$$\begin{bmatrix} 0 & 0 & 0 & 0 \\ 0 & \alpha\alpha^* & \alpha\beta^* & \alpha\gamma^* \\ 0 & \beta\alpha^* & \beta\beta^* & \beta\gamma^* \\ 0 & \gamma\alpha^* & \gamma\beta^* & \gamma\gamma^* \end{bmatrix} \text{ and } \begin{bmatrix} 0 & 0 & 0 & 0 \\ 0 & \Pr(3\,s) & 0 & 0 \\ 0 & 0 & \Pr(4s) & 0 \\ 0 & 0 & 0 & \Pr(5\,s) \end{bmatrix}.$$

In this example, we can form the partition π of the figures according to the parity (odd or even) of the number of sides so $\pi = \{O, E\} = \{\{3\,h, 3\,s, 5\,s\}, \{4\,s\}\}$. The indit set of the partition is the equivalence relation $(O \times O) \cup (E \times E) \subseteq U \times U$. Since the superposition represented by $\text{In}(S \times S)$ is indefinite between the solid figures, we could construct a more definite state by making distinctions or 'measuring' it according to parity. To represent that making-more-definite or 'measuring' operation using matrix multiplication, we need the projection matrices to the blocks of the parity partition:

$$P_O = \text{In}(\Delta O) \text{ and } P_E = \text{In}(\Delta E).$$

Then the making-more-definite or 'measurement' operation can be represented as the sum of matrix products:

$$P_O \, \text{In}(S \times S) P_O + P_E \, \text{In}(S \times S) P_E =$$
$$\begin{bmatrix} 1&0&0&0 \\ 0&1&0&0 \\ 0&0&0&0 \\ 0&0&0&1 \end{bmatrix}\begin{bmatrix} 0&0&0&0 \\ 0&1&1&1 \\ 0&1&1&1 \\ 0&1&1&1 \end{bmatrix}\begin{bmatrix} 1&0&0&0 \\ 0&1&0&0 \\ 0&0&0&0 \\ 0&0&0&1 \end{bmatrix}$$
$$+ \begin{bmatrix} 0&0&0&0 \\ 0&0&0&0 \\ 0&0&1&0 \\ 0&0&0&0 \end{bmatrix}\begin{bmatrix} 0&0&0&0 \\ 0&1&1&1 \\ 0&1&1&1 \\ 0&1&1&1 \end{bmatrix}\begin{bmatrix} 0&0&0&0 \\ 0&0&0&0 \\ 0&0&1&0 \\ 0&0&0&0 \end{bmatrix} = \begin{bmatrix} 0&0&0&0 \\ 0&1&0&1 \\ 0&0&1&0 \\ 0&1&0&1 \end{bmatrix}.$$

This incidence matrix represents the 'mixed' state of the superposition $\{3\,s, 5\,s\}$ and the singleton state $\{4\,s\}$ which is the incidence matrix of the relation $(\{3\,s, 5\,s\} \times \{3\,s, 5\,s\}) \cup (\{4\,s\} \times \{4\,s\}) \subseteq U \times U$. This 'measurement' transformed the indefinite superposition state $\{\{3\,s, 4\,s, 5\,s\}\}$ on the set of solid figures into the more definite mixed state $\{\{3\,s, 5\,s\}, \{4\,s\}\}$. Recalling that the distinctions or dits are the ordered pairs of elements in different blocks, the ditset of $\{\{3\,s, 5\,s\}, \{4\,s\}\}$ has four new dits: $(3\,s, 4\,s)$, $(5\,s, 4\,s)$, $(4\,s, 3\,s)$, and $(4\,s, 5\,s)$ in comparison with the original superposition state $\{\{3\,s, 4\,s, 5\,s\}\}$. The 'measurement' process created 4 distinctions. It is a theorem that this number of dits created is the number of non-zero off-diagonal elements that are zeroed in the 'measurement' process:

$$\begin{bmatrix} 0&0&0&0 \\ 0&1&1&1 \\ 0&1&1&1 \\ 0&1&1&1 \end{bmatrix} \rightsquigarrow \begin{bmatrix} 0&0&0&0 \\ 0&1&0&1 \\ 0&0&1&0 \\ 0&1&0&1 \end{bmatrix}.$$

This result restated for 'classical' and quantum logical entropy (defined below) shows that logical entropy gives the precise increase in information-as-dits (or qudits) in the process of measurement, i.e., logical entropy measures measurement.

To force a distinction in the superposition $\{3\,s, 5\,s\}$, we can perform another 'measurement' by the partition $\sigma = \{\{3\,h, 3\,s\}, \{4\,s, 5\,s\}\}$ that distinguishes the 3-sided figures from the non-3-sided figures. The result of that operation is calculated in the same way as:

$$
P_3 \begin{bmatrix} 0\,0\,0\,0 \\ 0\,1\,0\,1 \\ 0\,0\,1\,0 \\ 0\,1\,0\,1 \end{bmatrix} P_3 + P_{\neg 3} \begin{bmatrix} 0\,0\,0\,0 \\ 0\,1\,0\,1 \\ 0\,0\,1\,0 \\ 0\,1\,0\,1 \end{bmatrix} P_{\neg 3}
$$

$$
= \begin{bmatrix} 1\,0\,0\,0 \\ 0\,1\,0\,0 \\ 0\,0\,0\,0 \\ 0\,0\,0\,0 \end{bmatrix} \begin{bmatrix} 0\,0\,0\,0 \\ 0\,1\,0\,1 \\ 0\,0\,1\,0 \\ 0\,1\,0\,1 \end{bmatrix} \begin{bmatrix} 1\,0\,0\,0 \\ 0\,1\,0\,0 \\ 0\,0\,0\,0 \\ 0\,0\,0\,0 \end{bmatrix} + \begin{bmatrix} 0\,0\,0\,0 \\ 0\,0\,0\,0 \\ 0\,0\,1\,0 \\ 0\,0\,0\,1 \end{bmatrix} \begin{bmatrix} 0\,0\,0\,0 \\ 0\,1\,0\,1 \\ 0\,0\,1\,0 \\ 0\,1\,0\,1 \end{bmatrix} \begin{bmatrix} 0\,0\,0\,0 \\ 0\,0\,0\,0 \\ 0\,0\,1\,0 \\ 0\,0\,0\,1 \end{bmatrix}
$$

$$
= \begin{bmatrix} 0\,0\,0\,0 \\ 0\,1\,0\,0 \\ 0\,0\,1\,0 \\ 0\,0\,0\,1 \end{bmatrix} = \text{In}(\Delta S).
$$

The new distinctions created in this 'measurement' process are $(3\,s, 5\,s)$ and $(5\,s, 3\,s)$, and that is the number of non-zero off-diagonal elements zeroed in the process.

This 'measurement' operation on incidence matrices prefigures projective measurement in QM described by the Lüders mixture operation on density matrices. The initial state $\text{In}(S \times S)$ prefigures the density matrix of the pure superposition state of three eigenstates $3\,s$, $4\,s$, and $5\,s$, and the final result $\text{In}(\Delta S)$ prefigures the diagonal density matrix representing the completely decomposed 'classical' mixture of those eigenstates. As will be seen below, dividing these incidence matrices through by their trace (sum of diagonal elements) yields density matrices.

2.2 The Quantitative Measure of Information on Partitions

Since partitions are the mathematical concept to represent distinctions and indistinctions (or definiteness and indefiniteness), there should be a quantitative measure (in the sense of measure theory) to quantify the notion of information-as-distinctions. Since U is a finite set and the set of distinctions of a partition π is finite, the obvious notion to measure distinctions is simply the cardinality of the set of distinctions $\text{dit}(\pi) \subseteq U \times U$ normalized by the size of $U \times U$. Hence the *logical entropy of a partition* $\pi = \{B_1, ..., B_m\}$ for equiprobable points in U is:

$$
h(\pi) = \frac{|\text{dit}(\pi)|}{|U \times U|} = \frac{|U \times U| - |\cup_j B_j \times B_j|}{|U \times U|} = 1 - \sum_j \frac{|B_j|^2}{|U|^2} = 1 - \sum_j \Pr(B_j)^2
$$

Table 2.3 Classical logical probability theory and 'classical' logical information theory

Logical probability theory	Logical information theory
Elements $u \in U$ finite	Dits $(u, u') \in U \times U$ finite
Subsets $S \subseteq U$	Ditsets $\mathrm{dit}(\pi) \subseteq U \times U$
$\Pr(S) = \frac{\lvert S \rvert}{\lvert U \rvert}$	$h(\pi) = \frac{\lvert \mathrm{dit}(\pi) \rvert}{\lvert U \times U \rvert}$
$\Pr(S) = \sum \{ p_j : u_j \in S \}$	$h(\pi) = \sum \{ p_j p_k : (u_j, u_k) \in \mathrm{dit}(\pi) \}$
$\Pr(S) = $ 1-draw prob. of S-its	$h(\pi) = $ 2-draw prob. of π-dits

where $\Pr(B_j) = \lvert B_j \rvert / \lvert U \rvert$ is the probability that a random draw from U gives an element of B_j (Ellerman 2009, 2021, 2022). When the points of U have the point probabilities p_i for $i = 1, ..., n$, then:

$$h(\pi) = 1 - \sum_j \Pr(B_j)^2$$

where $\Pr(B_j) = \sum_{u_i \in B_j} p_i$. The logical entropy of π has an immediate interpretation; it is the probability that in two independent random draws from U, one will obtain a distinction of π. In view of the its-dits duality, this is just the two-draw version of $\Pr(S)$ which is the probability that in one draw from U, one will obtain an element of S. Since the indiscrete partition makes no distinctions, its logical entropy is $h(\mathbf{0}_U) = 0$. The discrete partition makes all possible distinctions so its logical entropy is $h(\mathbf{1}_U) = 1 - \sum_{i=1}^{n} p_i^2$,[5] which, in the equiprobable case, is $1 - \frac{1}{n}$, the probability that whatever was drawn on the first draw will not be drawn on the second draw.

The underlying duality of elements (its) and distinctions (dits) answers the question about the "size" of a partition. The size of a subset S is the number $\lvert S \rvert$ of elements or "its" in the subset and the size of a partition π is the number $\lvert \mathrm{dit}(\pi) \rvert$ of dits in the partition. The lattice of partitions is isomorphic to the lattice of ditsets partially ordered by inclusion (since refinement is just inclusion of ditsets), and the normalized size of subsets and ditsets (equiprobable case) gives the notions of probability $\Pr(S) = \frac{\lvert S \rvert}{\lvert U \rvert}$ and logical entropy $h(\pi) = \frac{\lvert \mathrm{dit}(\pi) \rvert}{\lvert U \times U \rvert}$–as summarized in Table 2.3.

When the point probabilities on U are given by the probability distribution $p = (p_1, ..., p_n)$, then the logical entropy $h(\pi)$ is the product probability measure $p \times p$ (defined on $U \times U$) of the ditset $\mathrm{dit}(\pi) \subseteq U \times U$. The logical entropy is thus the value of a measure in the sense of measure theory (unlike Shannon entropy (Ellerman 2021)). Information is defined as distinctions and the Shannon entropy is also a distinction-based definition. The Shannon entropy of a partition is the minimum average number of (equally weighted) binary partitions (bits) it takes to make all the distinctions of the partition. The compound notions of logical entropy are naturally defined in the usual Venn diagram manner as illustrated in Fig. 2.3 which includes the conditional logical entropy $h(\sigma \mid \pi)$ (the measure of the distinctions in σ that were not in π) and the mutual logical information $m(\pi, \sigma)$ (the measure of the distinctions common to π and σ).

---

[5] Brukner and Zeilinger have suggested this same formula (Brukner and Zeilinger 1999, 2003).

Fig. 2.3 $h(\pi \vee \sigma) =$
$h(\pi) + h(\sigma) - m(\pi, \sigma) =$
$h(\pi|\sigma) + h(\sigma|\pi) + m(\pi, \sigma)$

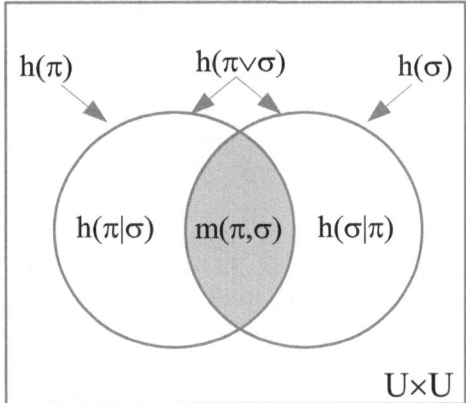

The joint notion of logical entropy that we will make later use of in the analysis of quantum measurement is the logical entropy $h(\pi \vee \sigma)$ of the join $\pi \vee \sigma$ which is the probability measure $p \times p$ on $\mathrm{dit}(\pi \vee \sigma) = \mathrm{dit}(\pi) \cup \mathrm{dit}(\sigma)$. That is the operation at the set level that prefigures projective measurement in quantum mechanics.

In spite of not being a measure in the sense of measure theory, the compound notions of Shannon entropy were defined to also satisfy those Venn diagram relationships. That is explained because there is a non-linear monotonic dit-to-bit transformation that transforms all the compound formulas of logical entropy into the corresponding formulas for Shannon entropy, and that transform preserves the Venn diagram relationships (Ellerman 2021).

2.3   Quantum States: Partitions, Density Matrices, and the Born Rule

Our thesis is that the mathematics of quantum mechanics is the mathematics of partitions linearized to vector (particularly, Hilbert) spaces. Partitions as a purely set concept (i.e., without point probabilities) will be linearized by the Yoga of Linearization to be described below. But it is convenient first to show how partitions (or equivalently equivalence relations) with point probabilities can be reformulated with set-level density matrices which then correlate with or prefigure the quantum-level density matrices representing quantum states.

A partition on a set U can be defined in the usual terms of a set of non-empty disjoint subsets $\pi = \{B_j\}_{j=1}^m$ whose union is the whole set. But a partition π can be equally well specified by its indit set $\mathrm{indit}(\pi) = \cup_{j=1}^m B_j \times B_j$ (the corresponding equivalence relation) or by the complementary ditset $\mathrm{dit}(\pi) = U \times U - \mathrm{indit}(\pi)$. The vector space notion that can represent the indit set of a partition (by its non-zero

entries), the ditset of a partition by its off-diagonal zero elements, and the point probabilities (by its diagonal entries) is the notion of a density matrix. In quantum mechanics, the state of a system can be represented by state vectors or by density matrices
(Nielsen and Chuang 2000, p. 102; Weinberg 2014). A partition-with-probabilities
correlates with or prefigures a quantum state expressed by a density matrix.

Consider a partition $\pi = \{B_1, ..., B_m\}$ on $U = \{u_1, ..., u_n\}$ with (always positive)
point probabilities $p = (p_1, ..., p_n)$. To reformulate the partition as a density matrix,
the blocks represent superposition subsets so let's denote such a subset as ΣB_j in
distinction from the discrete subset B_j. Then an $n \times n$ density matrix $\rho(\Sigma B_j)$ can
be defined for each block $\Sigma B_j \in \pi$ as follows:

$$\rho(\Sigma B_j)_{ik} = \{ \begin{array}{l} \frac{\sqrt{p_i p_k}}{\Pr(B_j)} \text{ if } (u_i, u_k) \in B_j \times B_j \\ 0 \text{ otherwise} \end{array} ..$$

where $\Pr(B_j) = \sum_{u_i \in B_j} p_i$. In the equiprobable case, this is just the incidence matrix
$In(B_j \times B_j)$ normalized by its trace (sum of diagonal elements). In Dirac notation,
if we let $\langle u_i | b_j \rangle = \frac{\sqrt{p_i}}{\sqrt{\Pr(B_j)}}$ if $u_i \in B_j$, else 0 (where $|u_i\rangle$ is the unit column vector
with a one in the ith place and zeros elsewhere). Then $\rho(\Sigma B_j)$ is the outer product
$\rho(\Sigma B_j) = |b_j\rangle \langle b_j|$ which is a pure state density matrix.

It is important to pause at this point to see the origin of the Born Rule at the set
level in the notion of a superposition set ΣS (modeled with the incidence matrix
$In(S \times S)$) as opposed to usual notion of a discrete subset S (modeled with the
incidence matrix $In(\Delta S)$) (Ellerman 2020). Using $\Sigma S = \Sigma B_j$ for a non-empty sub-
set $S \subseteq U$ considered as a superposition subset, then the $|b_j\rangle$ formula becomes:
$|s\rangle = \sum_{u_i \in S} \frac{\sqrt{p_i}}{\sqrt{\Pr(S)}} |u_i\rangle$. Thus $|s\rangle$ is just the $n \times 1$ column vector with the ith entry
$\frac{\sqrt{p_i}}{\sqrt{\Pr(S)}} \chi_S(u_i)$. Then the density matrix $\rho(\Sigma S)$ defined by the superposition subset
ΣS is the projection matrix (and outer product): $\rho(\Sigma S) = |s\rangle \langle s|$. As a pure density
matrix, $\rho(\Sigma S)$, it has one eigenvalue of 1 with the rest being zeros.

Given only a pure state density matrix $\rho(\Sigma S)$, the $|s\rangle$ is recovered as the normal-
ized eigenvector associated with the eigenvalue of 1 since $\rho(\Sigma S) = |s\rangle \langle s|$ follows
by the spectral decomposition of $\rho(\Sigma S)$ as a Hermitian matrix. These vectors $|s\rangle$
thus play the role of the state vectors at the set level and their coefficients are the
'amplitudes' (as opposed to probabilities) at the set level. There are no waves in sight
at the set level so those state vectors will *not* be called "wavefunctions."

The important thing to notice is that the (real) " amplitudes" $\frac{\sqrt{p_i}}{\sqrt{\Pr(S)}} \chi_S(u_i)$ appear
already at the set level once we treat superposition subsets using incidence and density
matrices. And the Born Rule appears at the set level to compute probabilities as the
amplitudes squared, e.g., the probability of u_i given the superposition subset ΣS:

$$\langle u_i | s \rangle^2 = (\frac{\sqrt{p_i}}{\sqrt{\Pr(S)}} \chi_S(u_i))^2 = \frac{p_i}{\Pr(S)} \chi_S(u_i) = \frac{\Pr(\{u_i\} \cap S)}{\Pr(S)} = \Pr(u_i | S).$$
$$\text{Set version of the Born Rule}$$

Then, as per our basic thesis, the quantum Born Rule is just the Hilbert space version
of the set level Born Rule (so it would use the absolute squares rather than the ordinary

squares). In both cases, it is a consequence of superposition. In the classical case of a fully definite mixed state (with probabilities given in the diagonal density matrix) where there is no superposition, then we are back in classical probability theory with no Born rule.

Returning to the density matrices $\rho(\Sigma B_j) = |b_j\rangle\langle b_j|$ which treat the blocks as superposition sets, they are combined to form the density matrix $\rho(\pi)$ representing the partition π as follows:

$$\rho(\pi) = \sum_{j=1}^{m} \Pr(B_j)\rho(\Sigma B_j) = \sum_{j=1}^{m} \Pr(B_j)|b_j\rangle\langle b_j|$$

so the entries are:

$$\rho(\pi)_{ik} = \begin{cases} \sqrt{p_i\, p_k} \text{ if } (u_i, u_k) \in \text{indit}(\pi) \\ 0 \text{ otherwise} \end{cases}.$$

Thus the indits of π are represented by the non-zero entries of $\rho(\pi)$, the dits of π are represented by the zero off-diagonal entries, the disjoint blocks of π are represented by the orthonormal vectors $|b_j\rangle$, and the point probabilities are given by the diagonal entries.

These density matrices over the reals are symmetric and have trace (sum of diagonal elements) equal to 1 since the diagonal elements are $\sqrt{p_i\, p_i} = p_i$ for $i = 1, ..., n$. The set notion of sum correlates with the matrix notion of trace. The probability p_i of an element u_i is recovered as $\text{tr}[P_{\{u_i\}}\rho(\pi)] = \sqrt{p_i}\sqrt{p_i} = p_i$ [where $P_{\{u_i\}}$ is the diagonal projection matrix with entries $\chi_{\{u_i\}}()$]. In the matrix language, the probability of a subset $S \subseteq U$ is: $\Pr(S) = \sum_{u_i \in S} p_i = \text{tr}[P_S \rho(\pi)]$ where P_S is the diagonal matrix with diagonal elements $\chi_S(u_i)$.

In general, density matrices can be defined as the probabilistic mixture $\rho = \sum_j w_j|\phi_j\rangle\langle\phi_j|$ of the projectors determined by normalized pure states $\{\phi_j\}$ and the $\{w_j\}$ form a probability distribution (Fano 1957). A general density matrix ρ is positive Hermitian, so the spectral decomposition gives orthonormal eigenvectors $\{|u_i\rangle\}_{i=1}^{n}$ with non-negative eigenvalues $\{\lambda_i\}_{i=1}^{n}$ that sum to 1. Hence $\rho = \sum_{i=1}^{n} \lambda_i|u_i\rangle\langle u_i|$ which is the general Hilbert space density matrix prefigured by the set level $\rho(\pi) = \sum_{j=1}^{m} \Pr(B_j)|b_j\rangle\langle b_j|$ where $\langle u_i|u_k\rangle = \delta_{ik}$ just as $\langle b_i|b_k\rangle = \delta_{ik}$.

Lemma 2.1 *The m non-zero eigenvalues of the density matrices $\rho(\pi)$ are the block probabilities $\Pr(B_j)$, with zeros filling out the remaining $n - m$ eigenvalues.*

Proof The eigenvalue equation is $\rho(\pi)v = \rho_j v$ or $\sum_{k=1}^{m} \Pr(B_k)|b_k\rangle\langle b_k|v = \rho_j v$ so for $v = |b_j\rangle$, we have: $\sum_k \Pr(B_k)|b_k\rangle\langle b_k|b_j\rangle = \rho_j|b_j\rangle$ and since $\langle b_k|b_j\rangle = \delta_{jk}$, we have $\rho(\pi)|b_j\rangle = \Pr(B_j)|b_j\rangle$ so $\rho_j = \Pr(B_j)$ are eigenvalues for $j = 1, ..., m$, with 0 as the remaining $n - m$ eigenvalues since the eigenvalues of a density matrix have to be non-negative and sum to one (Hughes 1989, p. 138). $\square$

The non-zero off-diagonal elements indicate the indistinctions of π where elements u_i and u_k 'cohere' together in the same block of the partition π and are called "coherences" in the case of quantum density matrices (Cohen-Tannoudji et al. 2005, p. 303).

[T]he off-diagonal terms of a density matrix … are often called quantum coherences because
they are responsible for the interference effects typical of quantum mechanics that are absent
in classical dynamics. (Auletta et al. 2009, p. 177).

Thus at the logical level, partition indistinctions (i.e., two different elements in the same block of a partition) prefigure quantum coherences of a superposition, and density matrices (as opposed to wave functions) give that direct representation of indistinctions of a partition as the non-zero elements of the corresponding density matrix. Thus density matrices (with non-zero off-diagonal elements) directly represent the indefiniteness interpretation of superposition (as opposed to the wave-addition notion of superposition).

The restatement of set partition π concepts in terms of the set level density matrix $\rho(\pi)$ is summarized in Table 2.4. The partition is $\pi = \{B_1, ..., B_m\}$ on $U = \{u_1, ..., u_n\}$ with the positive point probabilities $p = \{p_1, ..., p_n\}$. Many of our tables are part of the translation dictionary between the set-level math of partitions and the Hilbert space math of QM to illustrate our basic thesis. Table 2.4 is not that type of table; it translates the partition math concepts into the equivalent density matrix form to set up the translation dictionary of the next table, Table 2.5.

This correspondence can be illustrated by considering the partition $\pi = \{\{a, c\}, \{b\}\}$ on $U = \{a, b, c\}$ with point probabilities $p = (p_a, p_b, p_c)$. The rows and columns of the density matrix are labeled alphabetically. The trivial indits of π

Table 2.4 Density matrix translation of partition concepts

Set concept with probabilities	Set level density matrix concept
Partition π with point probs. p	Density matrix $\rho(\pi) = \sum_{j=1}^{m} \Pr(B_j)\lvert b_j\rangle\langle b_j\rvert$
Point probabilities $\{p_1, ..., p_n\}$	Value of diagonal entries of $\rho(\pi)$
Trivial indits (u_i, u_i) of π	Diagonal entries of $\rho(\pi)$
Non-trivial indits of π	Non-zero off-diagonal entries of $\rho(\pi)$
Dits of π	Zero entries of $\rho(\pi)$
Sum $\Pr(B_j) = \sum_{u_i \in B_j} p_i$	Trace $\mathrm{tr}[P_{B_j}\rho(\pi)]$
Block probabilities $\Pr(B_j)$ in π	Eigenvalues $\neq 0$ of $\rho(\pi)$
Disjoint blocks in π	Orthog. eigenvectors in $\rho(\pi)$ eigenvalues $\neq 0$
Block prob. 1 of U in $\mathbf{0}_U = \{U\}$	Non-zero eigenvalue of 1 for $\rho(\mathbf{0}_U)$

Table 2.5 Set level and quantum state density matrices

Set-level density matrix concept	Hilbert space density matrix concept
Density matrix $\rho(\pi)$	Quantum state density matrix ρ
Set-level Born Rule	Quantum Born Rule
Non-zero off-diagonal entries	Non-zero off-diagonal coherences of ρ
$\{\Pr(B_j)\}_{j=1}^{m}$ = non-0 eigenvalues	$\{\lambda_j\}_{j=1}^{m}$ eigenvalues, $\lambda_j \geq 0$, $\sum_{j=1}^{m} \lambda_j = 1$
$\rho(\pi) = \sum_{j=1}^{m} \Pr(B_j)\lvert b_j\rangle\langle b_j\rvert$	$\rho = \sum_{j=1}^{m} \lambda_j \lvert u_j\rangle\langle u_j\rvert$
Eigenvalue of 1 for $\rho(\mathbf{0}_U)$	Pure state ρ has eigenvalue 1

are: $\{\,(a, a), (b, b), (c, c)\}$ with the non-zero probabilities $p_a = \sqrt{p_a}\sqrt{p_a}$, p_b, and p_c, the diagonal elements of $\rho(\pi)$. The non-trivial indits of π are: $\{(a, c), (c, a)\}$ with the entries $\sqrt{p_a p_c}$ and $\sqrt{p_c p_a}$ as the non-zero off-diagonal elements of $\rho(\pi)$. The dits of π are: $\{(a, b), (b, c), (b, a), (c, b)\}$ which correspond to the zero entries of the density matrix:

$$
\rho(\{\{a, c\}, \{b\}\}) = \begin{bmatrix} p_a & 0 & \sqrt{p_a p_c} \\ 0 & p_b & 0 \\ \sqrt{p_c p_a} & 0 & p_c \end{bmatrix} = \{ \begin{array}{l} \text{Non-0 entries = indistinctions} \\ \text{= coherences, and.} \\ \text{Zero entries = distinctions} \end{array}
$$

The only coherences in $\rho(\pi)$ are the two entries $\sqrt{p_a p_c}$ since the only non-singleton block (i.e., "superposition states") in $\pi = \{\{a, c\}, \{b\}\}$ is the block $\{a, c\}$.

For a numerical example, take $p_a = \frac{1}{4}$, $p_b = \frac{1}{3}$, and $p_c = \frac{5}{12}$ so that:

$$
\rho(\{\{a, c\}, \{b\}\}) = \frac{2}{3} \begin{bmatrix} \frac{3}{8} & 0 & \frac{\sqrt{15}}{8} \\ 0 & 0 & 0 \\ \frac{\sqrt{15}}{8} & 0 & \frac{5}{8} \end{bmatrix} + \frac{1}{3} \begin{bmatrix} 0 & 0 & 0 \\ 0 & 1 & 0 \\ 0 & 0 & 0 \end{bmatrix} = \begin{bmatrix} \frac{1}{4} & 0 & \frac{\sqrt{15}}{12} \\ 0 & \frac{1}{3} & 0 \\ \frac{\sqrt{15}}{12} & 0 & \frac{5}{12} \end{bmatrix}.
$$

Then the disjoint partition blocks with their block probabilities become the normalized *orthogonal* eigenvectors with non-negative eigenvalues that sum to one:

$$
\{ \begin{bmatrix} \frac{\sqrt{3}}{2\sqrt{2}} \\ 0 \\ \frac{\sqrt{5}}{2\sqrt{2}} \end{bmatrix} \} \leftrightarrow \frac{2}{3}, \{ \begin{bmatrix} 0 \\ 1 \\ 0 \end{bmatrix} \} \leftrightarrow \frac{1}{3}.
$$

Then, as a Hermitian matrix, the density matrix $\rho(\{\{a, c\}, \{b\}\})$ can be expressed as the sum of the non-zero eigenvalues times the projections to their eigenspaces.

$$
\rho(\{\{a, c\}, \{b\}\}) = \frac{2}{3} \begin{bmatrix} \frac{\sqrt{3}}{2\sqrt{2}} \\ 0 \\ \frac{\sqrt{5}}{2\sqrt{2}} \end{bmatrix} \begin{bmatrix} \frac{\sqrt{3}}{2\sqrt{2}} & 0 & \frac{\sqrt{5}}{2\sqrt{2}} \end{bmatrix} + \frac{1}{3} \begin{bmatrix} 0 \\ 1 \\ 0 \end{bmatrix} \begin{bmatrix} 0 & 1 & 0 \end{bmatrix} = \begin{bmatrix} \frac{1}{4} & 0 & \frac{\sqrt{15}}{12} \\ 0 & \frac{1}{3} & 0 \\ \frac{\sqrt{15}}{12} & 0 & \frac{5}{12} \end{bmatrix}. \checkmark
$$

All of this translation of partition concepts into density matrices at the set level has a correlate or counterpart in the case of arbitrary density matrices in QM. A density matrix ρ is not only a Hermitian matrix but is a positive matrix so it has a spectral decomposition with a basis set of orthonormal eigenvectors $\{|u_j\rangle\}$ and non-negative eigenvalues $\{\lambda_j\}$ that sum to one. Hence just like $\rho(\{\{a, c\}, \{b\}\})$, an arbitrary density matrix of QM can be expressed as a probabilistic sum of projectors formed from orthonormal vectors:

$$
\rho = \sum_{i=1}^{n} \lambda_i |u_i\rangle\langle u_i|.
$$

Thus by using the intermediate step of representing a set partition π on U with point probabilities as a density matrix $\rho(\pi) = \sum_{j=1}^{m} \Pr(B_j)|b_j\rangle\langle b_j|$, we see how a partition with point probabilities has its correlate in the mathematics of QM as a density matrix $\rho = \sum_{i=1}^{n} \lambda_i |u_i\rangle\langle u_i|$. That is an example of how the math of partitions is the skeletal version that prefigures the corresponding Hilbert space math of QM. The

Hilbert space notion of a quantum state represented by a density matrix is prefigured by a partition on a set with point probabilities. These results are summarized in the following Table 2.5.

2.4 Logical Entropy with Density Matrices

In the formula for logical entropy $h(\pi) = 1 - \sum_j \Pr(B_j)^2$, the density matrix replaces the block probabilities and the trace replaces the summation to give the same result:

$$h(\pi) = 1 - \sum_j \Pr(B_j)^2 = 1 - \text{tr}[\rho(\pi)^2].$$

These real-valued density matrices encapsulate the 'classical' treatment of the mathematics of partitions that prefigures the quantum treatment. Two of these classical results translate directly into the corresponding results in quantum mechanics.

The first result is that the projective measurement in QM is classically just the join of partitions at the skeletal level of set partitions. We start with the partition π with point probabilities expressed by the density matrix $\rho(\pi)$ and then we think of the partition $\sigma = \{C_1, ..., C_{m'}\}$ on U as being the inverse-image g^{-1} of a numerical attribute or 'observable' $g : U \to \mathbb{R}$. In QM, the effect of a projective measurement on a density matrix ρ is given by the *Lüders mixture operation* (Lüders 1951; Furry 1966, p. 16; Auletta et al. 2009, p. 279) or *Lüders transformer* (Busch 1995, p. 38). In the skeletal version, for each block $C_j \in \sigma$, let P_{C_j} be the projection matrix that is a diagonal matrix with the diagonal elements given by the characteristic function $\chi_{C_j} : U \to 2 = \{0, 1\}$ of C_j. Then the Lüders mixture operation transforms the density matrix $\rho(\pi)$ into the density matrix $\hat{\rho}(\pi)$ according to the formula:

$$\hat{\rho}(\pi) = \sum_{C_j \in \sigma} P_{C_j} \rho(\pi) P_{C_j}.$$
Lüders Mixture Operation

Theorem 2.1 $\hat{\rho}(\pi) = \rho(\pi \vee g^{-1})$.

Proof A non-zero entry in $\rho(\pi)$ has the form $\rho(\pi)_{ik} = \sqrt{p_i p_k}$ iff there is some block $B \in \pi$ such that $(u_i, u_k) \in B \times B$, i.e., if $u_i, u_k \in B$. The matrix operation $P_{C_j}\rho(\pi)$ will preserve the entry $\sqrt{p_i p_k}$ if $u_i \in C_j$, otherwise the entry is zeroed. And if the entry was preserved, then the further matrix operation $(P_{C_j}\rho(\pi))P_{C_j}$ will preserve the entry $\sqrt{p_i p_k}$ if $u_k \in C_j$, otherwise it is zeroed. Hence the entries $\sqrt{p_i p_k}$ in $\rho(\pi)$ that are preserved in $P_{C_j}\rho(\pi)P_{C_j}$ are the entries where both $u_i, u_k \in B$ for some $B \in \pi$ and $u_i, u_k \in C_j$. Recall that $\text{dit}(\pi \vee \sigma) = \text{dit}(\pi) \cup \text{dit}(\sigma)$ so applying DeMorgan's law, $\text{indit}(\pi \vee \sigma) = \text{dit}(\pi \vee \sigma)^c = \text{dit}(\pi)^c \cap \text{dit}(\sigma)^c = \text{indit}(\pi) \cap \text{indit}(\sigma)$. Thus the join of partitions is just the partition corresponding to the equivalence relation resulting from intersecting two equivalence relations (indit sets). These are the entries in $\rho(\pi \vee \sigma)$ corresponding to the blocks $B \cap C_j$ for some $B \in \pi$, so summing over $C_j \in \sigma$ gives: $\sum_{C_j \in \sigma} P_{C_j} \rho(\pi) P_{C_j} = \hat{\rho}(\pi) = \rho(\pi \vee \sigma) = \rho(\pi \vee g^{-1})$. $\square$

Our theme is that the vector space mathematics of QM is prefigured at the logical level by the mathematics of partitions on sets. The above Theorem shows that the standard partition operation of join of partitions (or intersection of equivalence relations) is essentially the set version of the projective measurement operation (Lüders mixture operation) in QM. Note that partitions have two separate roles in this set-based example; π with its point probabilities represents the state being measured and σ represents the numerical attribute (or observable) being measured on that state. The join operation joins the two partitions and creates more distinctions since $\text{dit}(\pi \vee \sigma) = \text{dit}(\pi) \cup \text{dit}(\sigma)$ or, in terms of equivalence relations, $\text{indit}(\pi \vee \sigma) = \text{indit}(\pi) \cap \text{indit}(\sigma)$ so the only indistinctions that 'survive' in the join are the indistinctions of both partitions. Since $\rho(\pi)$ represented the quantum state, the quantum notion of becoming is represented by the jump $\rho(\pi) \rightsquigarrow \rho(\pi \vee \sigma)$.

If $\pi \vee \sigma = \mathbf{1}_U$, then the measurement is said to be *non-degenerate* or *maximal* and $\rho(\mathbf{1}_U)$ is a diagonal matrix in the eigenstates of the observable. In the case $\pi = \mathbf{0}_U$ and $\mathbf{0}_U \vee \sigma = \sigma \neq \mathbf{1}_U$, then other partitions on the same set (i.e., other commuting observables in QM) are needed until we have $\mathbf{0}_U \vee \sigma \vee ... \vee \gamma = \sigma \vee ... \vee \gamma = \mathbf{1}_U$. Then $\sigma = g^{-1}, ..., \gamma = h^{-1}$ are the inverse-image partitions of numerical attributes $g, ..., h : U \to \mathbb{R}$ that would form a *complete set of compatible attributes* (CSCA) that prefigure a complete set of commuting observables (CSCO) in QM. The ordered set of $g, ..., h : U \to \mathbb{R}$ values for each $u_i \in U$ would uniquely characterize that element of U. That directly generalizes to the quantum case of CSCOs.

For the second result that generalizes to the quantum case, the off-diagonal non-zero entries in the density matrices represent the non-trivial indistinctions, so the distinctions that are created by joining σ with π will be indicated by those non-zero entries in $\rho(\pi)$ that are zeroed in $\hat{\rho}(\pi) = \rho(\pi \vee \sigma)$. Logical entropy measures information-as-distinctions so the non-zero off-diagonal entries that are zeroed, the indistinctions that become distinctions (i.e., the coherences that are decohered in the quantum case), will be measured by the increase in logical entropy (the increase in quantum logical entropy in the quantum case).

Theorem 2.2 (Set case of Measuring Measurement) *The sum of all the squares $p_i p_k$ of all the entries $\sqrt{p_i p_k}$ that were zeroed in the Lüders mixture operation that transforms $\rho(\pi)$ into $\hat{\rho}(\pi) = \sum_{C_j \in \sigma} P_{C_j} \rho(\pi) P_{C_j} = \rho(\pi \vee \sigma)$ is the increase in logical entropy: $h(\pi \vee \sigma) - h(\pi) = h(\sigma|\pi)$.*

Proof All the entries $\sqrt{p_i p_k}$ that got zeroed were for ordered pairs (u_i, u_k) that were indits of π but not indits of $\pi \vee \sigma$, i.e., $(u_i, u_k) \in \text{indit}(\pi) \cap \text{indit}(\pi \vee \sigma)^c = \text{dit}(\pi)^c \cap \text{dit}(\pi \vee \sigma) = \text{dit}(\pi \vee \sigma) - \text{dit}(\pi)$. The sum of products $p_i p_k$ for those pairs (u_i, u_k) is just the product probability measure on that set $\text{dit}(\pi \vee \sigma) - \text{dit}(\pi)$ which is $h(\pi \vee \sigma|\pi)$. And since $\text{dit}(\pi) \subseteq \text{dit}(\pi \vee \sigma)$, the measure on $\text{dit}(\pi \vee \sigma) - \text{dit}(\pi) = \text{dit}(\sigma) - \text{dit}(\pi)$ is $h(\pi \vee \sigma|\pi) = h(\pi \vee \sigma) - h(\pi) = h(\sigma|\pi)$ which is the information-as-distinctions that σ added to the information in π. □

Example: If the four states of $U = \{a, b, c, d\}$ were equiprobable, the real-valued density matrix of a particle in the mixed state represented by the partition $\pi = \{abc, d\}$ (in shorthand notation) is:

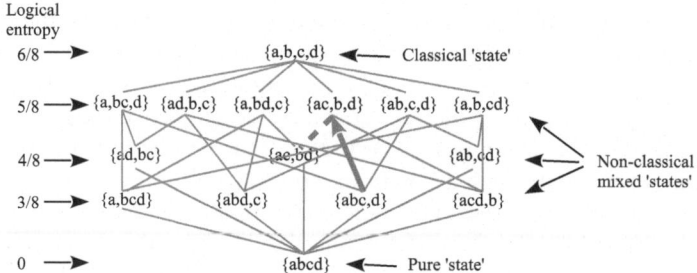

Fig. 2.4 $\pi \vee \sigma = \{abc, d\} \vee \{ac, bd\} = \{ac, b, d\}$ with the logical entropies of the different levels

$$\rho(\{abc, d\}) = \begin{bmatrix} \frac{1}{4} & \frac{1}{4} & \frac{1}{4} & 0 \\ \frac{1}{4} & \frac{1}{4} & \frac{1}{4} & 0 \\ \frac{1}{4} & \frac{1}{4} & \frac{1}{4} & 0 \\ 0 & 0 & 0 & \frac{1}{4} \end{bmatrix} = \{ \begin{array}{l} \text{Non-0 entries = indistinctions} \\ \text{= coherences, and} \\ \text{Zero entries = distinctions} \end{array} \quad .$$

The main partition operation representing going from an indefinite state or partition to a more definite (i.e., more refined) one is the *join* operation, for instance: $\{ac, bd\} \vee \{abc, d\} = \{ac, b, d\}$ as in Fig. 2.4 where $\sigma = \{ac, bd\}$. Since partitions with probabilities are the set-level versions of quantum states, we might carry over the quantum classification of states to such partitions. The set version of a pure state is the indiscrete partition $\mathbf{0}_U = \{\{a, b, c, d\}\}$ or $\{abcd\}$ in shorthand. The set version of a classical state, i.e., a state with no non-trivial indistinctions (no "coherences"), is the discrete partition $\mathbf{1}_U = \{\{a\}, \{b\}, \{c\}, \{d\}\}$ or $\{a, b, c, d\}$ in shorthand. And all the other partitions are the non-classical mixed states, i.e., with some non-trivial indistinctions (i.e., indistinctions other than the self-pairs (u_i, u_i) represented as the diagonal entries) and some distinctions, as illustrated in Fig. 2.4. The mixed states in partition lattice of Fig. 2.4 are all those that can be obtained by measurements on the bottom pure state $\{abcd\}$.

The pure math of the partition join is symmetrical, but the Lüders mixture operation is not symmetrical. In Fig. 2.4, $\pi = \{abc, d\}$ is the state being measured and $\sigma = \{ac, bd\}$ represents the observable. The measurement being represented is not maximal since there is still a non-singleton superposition state in the result $\pi \vee \sigma = \{ac, b, d\}$. The arrow from π to $\pi \vee \sigma$ is the set-level quantum jump from the less refined state π with logical entropy 3/8 to the more refined state with entropy 5/8.

Since logical entropy is a measure in the sense of measure theory (unlike Shannon entropy (Ellerman 2021)), the logical entropies can be presented as areas in a Venn diagram as in Fig. 2.5.

The QM-version of the partition join is a projective measurement described by the Lüders mixture operation. Since $\{ac, bd\}$ or without shorthand, $\{\{a, c\}, \{b, d\}\}$ is being joined to $\{\{a, b, c\}, \{d\}\}$, we need the projection matrices to $\{a, c\}$ and to $\{b, d\}$:

Fig. 2.5 Venn diagram with logical entropies for $\pi \vee \sigma$

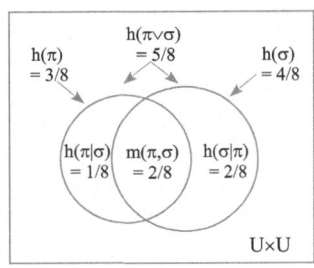

$$P_{\{a,c\}} = \begin{bmatrix} 1 & 0 & 0 & 0 \\ 0 & 0 & 0 & 0 \\ 0 & 0 & 1 & 0 \\ 0 & 0 & 0 & 0 \end{bmatrix} \text{ and } P_{\{b,d\}} = \begin{bmatrix} 0 & 0 & 0 & 0 \\ 0 & 1 & 0 & 0 \\ 0 & 0 & 0 & 0 \\ 0 & 0 & 0 & 1 \end{bmatrix}$$

which are the matrix versions of the set projection operations $\{a, c\} \cap ()$ and $\{b, c\} \cap$ $() : \wp(U) \to \wp(U)$. The Lüders mixture operation pre- and post-multiplies the pre-measurement density matrix $\rho(\{\{a, b, c\}, \{d\}\})$ by these two projection matrices and the result sums to yield the post-measurement mixture density matrix $\hat{\rho}$:

$$\hat{\rho} = P_{\{a,c\}}\rho(\{\{a, b, c\}, \{d\}\})P_{\{a,c\}} + P_{\{b,d\}}\rho(\{\{a, b, c\}, \{d\}\})P_{\{b,d\}};$$

$$\begin{bmatrix} 1 & 0 & 0 & 0 \\ 0 & 0 & 0 & 0 \\ 0 & 0 & 1 & 0 \\ 0 & 0 & 0 & 0 \end{bmatrix} \begin{bmatrix} \frac{1}{4} & \frac{1}{4} & \frac{1}{4} & 0 \\ \frac{1}{4} & \frac{1}{4} & \frac{1}{4} & 0 \\ \frac{1}{4} & \frac{1}{4} & \frac{1}{4} & 0 \\ 0 & 0 & 0 & \frac{1}{4} \end{bmatrix} \begin{bmatrix} 1 & 0 & 0 & 0 \\ 0 & 0 & 0 & 0 \\ 0 & 0 & 1 & 0 \\ 0 & 0 & 0 & 0 \end{bmatrix} = \begin{bmatrix} \frac{1}{4} & 0 & \frac{1}{4} & 0 \\ 0 & 0 & 0 & 0 \\ \frac{1}{4} & 0 & \frac{1}{4} & 0 \\ 0 & 0 & 0 & 0 \end{bmatrix} \text{ and;}$$

$$\begin{bmatrix} 0 & 0 & 0 & 0 \\ 0 & 1 & 0 & 0 \\ 0 & 0 & 0 & 0 \\ 0 & 0 & 0 & 1 \end{bmatrix} \begin{bmatrix} \frac{1}{4} & \frac{1}{4} & \frac{1}{4} & 0 \\ \frac{1}{4} & \frac{1}{4} & \frac{1}{4} & 0 \\ \frac{1}{4} & \frac{1}{4} & \frac{1}{4} & 0 \\ 0 & 0 & 0 & \frac{1}{4} \end{bmatrix} \begin{bmatrix} 0 & 0 & 0 & 0 \\ 0 & 1 & 0 & 0 \\ 0 & 0 & 0 & 0 \\ 0 & 0 & 0 & 1 \end{bmatrix} = \begin{bmatrix} 0 & 0 & 0 & 0 \\ 0 & \frac{1}{4} & 0 & 0 \\ 0 & 0 & 0 & 0 \\ 0 & 0 & 0 & \frac{1}{4} \end{bmatrix} \text{ and;}$$

$$\begin{bmatrix} \frac{1}{4} & 0 & \frac{1}{4} & 0 \\ 0 & 0 & 0 & 0 \\ \frac{1}{4} & 0 & \frac{1}{4} & 0 \\ 0 & 0 & 0 & 0 \end{bmatrix} + \begin{bmatrix} 0 & 0 & 0 & 0 \\ 0 & \frac{1}{4} & 0 & 0 \\ 0 & 0 & 0 & 0 \\ 0 & 0 & 0 & \frac{1}{4} \end{bmatrix} = \begin{bmatrix} \frac{1}{4} & 0 & \frac{1}{4} & 0 \\ 0 & \frac{1}{4} & 0 & 0 \\ \frac{1}{4} & 0 & \frac{1}{4} & 0 \\ 0 & 0 & 0 & \frac{1}{4} \end{bmatrix};$$

$$\hat{\rho} = \rho(\{\{a, c\}, \{b, d\}\} \vee \{\{a, b, c\}, \{d\}\}) = \rho(\{\{a, c\}, \{b\}, \{d\}\}).$$

The logical entropy of $\rho(\{\{a, b, c\}, \{d\}\})$ is

$$h(\rho(\{\{a, b, c\}, \{d\}\})) = 1 - \text{tr}[\rho(\{\{a, b, c\}, \{d\}\})^2] = 1 - \tfrac{10}{16} = \tfrac{3}{8}$$

and the logical entropy of $\hat{\rho} = \rho(\{\{a, c\}, \{b\}, \{d\}\})$ is

$$h(\rho(\{\{a, c\}, \{b\}, \{d\}\})) = 1 - \text{tr}[\rho(\{\{a, c\}, \{b\}, \{d\}\})^2] = 1 - \tfrac{3}{8} = \tfrac{5}{8}.$$

In the transition from $\rho(\{\{a, b, c\}, \{d\}\})$ to $\hat{\rho} = \rho(\{\{a, c\}, \{b\}, \{d\}\})$, there were four "coherences" of $\frac{1}{4}$ that were zeroed (four "decoherences" or dits created), so the sum of their squares is: $4 \times (\frac{1}{4})^2 = \frac{1}{4}$ which, by the Measuring Measurement Theorem, equals the increase in logical entropy; $\frac{5}{8} - \frac{3}{8} = \frac{1}{4}.\checkmark$

The post-measurement density matrix obtained by the Lüders mixture operation needs to be followed by the state reduction to one of the states in the mixture according to the eigenvalue of the observable that was returned. Assume that the eigenvalue returned was the (degenerate) one whose inverse-image was the state $\{a, c\}$. Then Lüders Rule (Hughes 1989, App. B) says that the final density matrix is the given term $P_{\{a,c\}}\rho(\{\{a, b, c\}, \{d\}\})P_{\{a,c\}}$ (from the Lüders mixture sum) divided by its trace. In this case, the final density matrix is for the pure superposition state $\{a, c\}$:

$$\frac{P_{\{a,c\}}\rho(\{\{a,b,c\},\{d\}\})P_{\{a,c\}}}{\mathrm{tr}[P_{\{a,c\}}\rho(\{\{a,b,c\},\{d\}\})P_{\{a,c\}}]} = \begin{bmatrix} \frac{1}{4} & 0 & \frac{1}{4} & 0 \\ 0 & 0 & 0 & 0 \\ \frac{1}{4} & 0 & \frac{1}{4} & 0 \\ 0 & 0 & 0 & 0 \end{bmatrix} \frac{1}{1/2} = \begin{bmatrix} \frac{1}{2} & 0 & \frac{1}{2} & 0 \\ 0 & 0 & 0 & 0 \\ \frac{1}{2} & 0 & \frac{1}{2} & 0 \\ 0 & 0 & 0 & 0 \end{bmatrix}.$$

If the other (degenerate) eigenvalue for $\{b, d\}$ had been returned, then the final density matrix would be for the (classical) mixture $\{\{b\}, \{d\}\}$:

$$\frac{P_{\{b,d\}}\rho(\{\{a,b,c\},\{d\}\})P_{\{b,d\}}}{\mathrm{tr}[P_{\{b,d\}}\rho(\{\{a,b,c\},\{d\}\})P_{\{b,d\}}]} = \begin{bmatrix} 0 & 0 & 0 & 0 \\ 0 & \frac{1}{4} & 0 & 0 \\ 0 & 0 & 0 & 0 \\ 0 & 0 & 0 & \frac{1}{4} \end{bmatrix} \frac{1}{1/2} = \begin{bmatrix} 0 & 0 & 0 & 0 \\ 0 & \frac{1}{2} & 0 & 0 \\ 0 & 0 & 0 & 0 \\ 0 & 0 & 0 & \frac{1}{2} \end{bmatrix}.$$

References

Auletta G, Fortunato M, Parisi G (2009) Quantum mechanics. Cambridge University Press, Cambridge, UK

Brukner Č, Zeilinger A (1999) Operationally invariant information in quantum measurements. Phys Rev Lett 83:3354–57

Brukner Č, Zeilinger A (2003) Information and fundamental elements of the structure of quantum theory. In: Castell L, Ischebeck O (eds) Time, quantum and information. Springer, Berlin, pp 323–354

Bunge M (2010) Matter and mind: a philosophical inquiry. Spring Publications, Dordrecht

Busch P, Grabowski M, Lahti PJ (1995) Operational quantum physics. Springer, Berlin

Chiatti L (2022) A neo-Aristotelic (maybe) approach to quantum reality. Biocosmology - Neo-Aristotelism 12:309–325

Cohen-Tannoudji C, Diu B, Laloë F (2005) Quantum mechanics: volumes 1 and 2. Wiley, New York

Del Santo F, Gisin N (2023) Potentiality realism: a realistic and indeterministic physics based on propensities. arXiv:2305.02429v2 [quant-ph]

deRonde C (2018) Quantum superpositions and the representation of physical reality beyond measurement outcomes and mathematical structures. Found Sci 23:621–648

Ellerman D (2009) Counting distinctions: on the conceptual foundations of Shannon's information theory. Synthese 168:119–149. https://doi.org/10.1007/s11229-008-9333-7

Ellerman D (2020) Probability theory with superposition events: a classical generalization in the direction of quantum mechanics. https://arxiv.org/abs/2006.09918

Ellerman D (2021) New foundations for information theory: logical entropy and Shannon entropy. Springer Nature, Cham, Switzerland

Ellerman D (2022) Introduction to logical entropy and its relationship to Shannon entropy. 4Open Special Issue: Logical Entropy 5:1–33. https://doi.org/10.1051/fopen/2021004

Fano U (1957) Description of states in quantum mechanics by density matrix and operator techniques. Rev Mod Phys 29:74–93

Feyerabend P (1983) Problems of microphysics. In: Colodny RG (ed) Frontiers of science and philosophy. University Press of America, Lanham, MD, pp 189–283

Fleming GN (1992) The actualization of potentialities in contemporary quantum theory. J Specul Philosop 6:259–276

Furry WH (1966) Some aspects of the quantum theory of measurement. In: Brittin WE (ed) Lectures in theoretical physics, vol 8A. University of Colorado Press, Boulder, CO, pp 1–64

Gisin N (2014) Quantum chance: nonlocality. Springer, Cham

Halmos PR (1974) Naive set theory. Springer Science+Business Media, New York

Hawkins D (1964) The language of nature: an essay in the philosophy of science. Anchor Books, Garden City, NJ

Heisenberg W (1962) Physics and philosophy: the revolution in modern science. Harper Torchbooks, New York

Herbert N (1985) Quantum reality: beyond the new physics. Anchor Books, New York

Hughes RIG (1989) The structure and interpretation of quantum mechanics. Harvard University Press, Cambridge

Jaeger G (2014) Quantum objects: non-local correlation. Causality and objective indefiniteness in the quantum world. Springer, Heidelberg

Jaeger G (2017) Quantum potentiality revisited. PhilTransRSocA 375:20160390. http://dx.doi.org/10.1098/rsta.2016.0390

Karakostas V (2007) Nonseparability, potentiality, and the context-dependence of quantum objects. J Gen Philos Sci 38:279–297

Kastner RE (2013) The transactional interpretation of quantum mechanics: the reality of possibility. Cambridge University Press, New York

Kastner RE (2015) Understanding our unseen reality: solving quantum riddles. Imperial College Press, London

Kastner RE, Kauffman S, Epperson M (2018) Taking Heisenberg's Potentia seriously. Int J Quantum Found 4:158–172

Kožnjak B (2007) Möglichkeit. Wirklichkeit Und Quantenmechanik. Prolegomena 6(2):223–52

Levy-Leblond JM, Balibar F (1990) Quantics: rudiments of quantum physics. North-Holland, Amsterdam

Lüders G (1951) Über die Zustandsänderung durch Meßprozeß. Annalen der Physik 8:322–328

Margenau H (1954) Advantages and disadvantages of various interpretations of the quantum theory. Phys Today 7:6–13

Mittelstaedt P (1998) The constitution of objects in Kant's philosophy and in modern physics. In: Castellani E (ed) Interpreting bodies: classical and quantum objects in modern physics. Princeton University Press, Princeton, pp 168–180

Nielsen M, Chuang I (2000) Quantum computation and quantum information. Cambridge University Press, Cambridge

Rohrlich F (1986) Reality and quantum mechanics. In: Greenberger DM (ed) New techniques and ideas in quantum measurement theory. New York Academy of Sciences, New York, pp 373–381

Rohrlich F (1987) From paradox to reality: our basic concepts of the physical world. Cambridge University Press, Cambridge, UK

Shimony A (1993) Search for a naturalistic world view: vol II natural science and metaphysics. Cambridge University Press, New York

Shimony A (1999) Philosophical and experimental perspectives on quantum physics. Epistemological and experimental perspectives on quantum physics: Vienna circle institute yearbook 7. Springer Science+Business Media, Dordrecht, pp 1–18

Strumia A (2021) A "Potency-Act" Interpretation of quantum physics. J Mod Phys 12:959–970. https://doi.org/10.4236/jmp.2021.127058

Weinberg S (2014) Quantum mechanics without state vectors. Phys Rev A 90:042102. https://doi.org/10.1103/PhysRevA.90.042102

Chapter 3
The Yoga of Linearization

About the mathematical tools, ...all of the physical quantities (time, energy, speed, position, etc.) continuous in a classical world, they are very well mathematically expressed by continuous variables and functions thereof. Heisenberg understood that these can no longer be the appropriate mathematical instruments and proposed to introduce operators as the mathematical representation of quantum mechanical quantities.

Gennaro Auletta and Shang-Yung Wang (2014)

Abstract This chapter gives the partitional treatment of the concept of a quantum observable. This is done using a bit of mathematical folklore that we call the "Yoga of Linearization." It is a method to transform set concepts into the corresponding vector space concepts. Starting with a real-valued numerical attribute $f : U \to \mathbb{R}$ on a set U (like weight, height, age, etc.), we apply the Yoga to obtain the notion of a real-valued operator on a vector space over a field containing $\mathbb{R}$. If U is taken as a basis set for a vector space over $\mathbb{C}$, then this procedure obtains a Hermitian operator, i.e., a quantum observable. But the Yoga also has an interesting intermediate step to transform set concepts into vector space concepts over the two element field $\mathbb{Z}_2$. This intermediate step gives a simplified pedagogical (or toy) model of QM over sets. Hence we develop that model to illustrate results in standard QM over $\mathbb{C}$ such as the double-slit experiment and Bell's Theorem.

Keywords Yoga of linearization · Quantum observables · Numerical attributes · Direct-sum decompositions · QM over sets · Double-slit experiment · Bell's theorem

3.1 From Set Concepts to Vector-Space Concepts

Our thesis is that the mathematics of QM is essentially the mathematics of partitions linearized to (Hilbert) vector spaces. There is a semi-algorithmic method–part of the folklore of mathematics–to linearize concepts using sets, e.g., set partitions (without

Table 3.1 Some initial applications of the Yoga

Set concept	Vector-space concept
Universe set U	U basis for space $[U] = \Bbbk^U = V$ over $\Bbbk$
Cardinality of the set U	Dimension of the space V
Subset S of the set U	Subspace $[S]$ of the space V

probabilities), to the corresponding concepts over vector spaces–a method that Gian-Carlo Rota might call a "yoga" (Rota 1997, p. 251). The idea is based on taking the vector space concept corresponding to the notion of a set as a basis set of the space. Then the yoga is:

> **The Yoga of Linearization**
> For any given set-concept, apply it to a basis set of a vector space and whatever is linearly generated is the corresponding vector space concept.

In applying the Yoga, we take U as being first a set and then a basis set of a vector space over a field $\Bbbk$. The associated functor is the free vector space or linearization functor from the category of *Sets* to the category of vector spaces $Vect_\Bbbk$ over $\Bbbk$ that takes U to $\Bbbk^U = V$. But the Yoga gives us a richer translation dictionary correlating set and vector-space concepts. The set concept of a subset S when applied to a basis set generates a subspace $[S]$. The cardinality of the U gives the dimension of the space generated by the basis set U as shown in Table 3.1.

If the vector space V has inner products, then we may assume U is an orthonormal basis of V.

We have already seen how the notion of a quantum state represented by a density matrix is prefigured by a partition with point probabilities. Now we deal with the other basic tool in QM mathematics, the concept of an observable (where there are no point probabilities in U).

3.2   Quantum Observables: Partitions and Direct-Sum Decompositions

Sometimes more than one vector space concept can be generated so a choice must be made. Consider the set concept of a numerical attribute $f : U \to \mathbb{R}$ taking values in the real numbers $\mathbb{R}$. Taking U as a basis set for a vector space V over $\mathbb{C}$, it defines both a linear functional $\hat{f} : V \to \mathbb{R}$ and a diagonalizable linear operator: $F : V \to V$ generated by $Fu_i = f(u_i) u_i$ on the basis vectors $u_i \in U$. The numerical attribute $f : U \to \mathbb{R}$ is recovered from the linear operator as the eigenvalue function assigning eigenvalues to the eigenvector basis set U.

Table 3.2 Operators corresponding to numerical attributes

Set concept	Vector-space concept
$f : U \to \mathbb{R}$	$F : V \to V$ by $F u_i = f(u_i) u_i$
$g : U \to \mathbb{R}$	$G : V \to V$ by $G u_i = g(u_i) u_i$
f, g on same set U	F, G comm. basis U of sim. eigenvectors

For the purposes of extending the mathematics of set partitions to vector spaces to obtain the mathematics of QM, it is the operator that is used in the Yoga, not the linear functional. Operators allow the discreteness so characteristic of quantum mechanics (and the underlying numerical attributes $f : U \to \mathbb{R}$ embody no notion of continuity).

The elements of the basis set U are a basis of eigenvectors and the values of the numerical attribute are the eigenvalues of the (always diagonalizable) operator F. Two numerical attributes defined on the *same* set U would generate two linear operators that commute with the basis set U as a basis of simultaneous eigenvectors as illustrated in Table 3.2.

The inverse-image of $f : U \to \mathbb{R}$ is a partition f^{-1} on U where each block $f^{-1}(r)$ is associated with a distinct $r \in f(U)$ a subset of $\mathbb{R}$. What is the vector-space version of a set partition (with no point probabilities)? The Yoga requires a choice.

If we had taken the vector-space analog of $f : U \to \mathbb{R}$ as the linear functional $\hat{f} : V \to \mathbb{R}$, then the corresponding vector space concept would be the inverse-image of the functional which is a special type of *set* partition on V called a *commuting* or *permutable partition* which was much studied by Gian-Carlo Rota and colleagues (Finberg et al. 1996) and earlier by Dubreil and Dubreil-Jacotin (1939) and Ore (1942).

But our choice is that the vector-space notion of a partition is a *direct-sum decomposition (DSD)* (Hoffman and Kunze 1961, p. 154). Each block in the set partition $f^{-1} = \left\{ f^{-1}(r) \right\}_{r \in f(U)}$ of the basis set U generates a subspace which is just the eigenspace V_r of the linear operator $F : V \to V$ (defined by $F u_i = f(u_i) u_i$) determined by the eigenvalue $r \in f(U)$, and the whole space is the direct sum: $V = \oplus_{r \in f(U)} V_r$.[1]

One characterization of a direct-sum decomposition $\{V_r\}_{r \in f(U)}$ is that every non-zero vector $v \in V$ can represented in exactly one way as the sum $v = \sum_{r \in f(U)} v_r$ of non-zero vectors $v_r \in V_r$. The vectors v_r are the projections $P_{V_r}(v)$ of v to the subspaces V_r. What happens if we put the Yoga in reverse and read that property of DSDs back into sets? A set partition is usually defined as a set of non-empty subsets $\{B_1, B_2, ..., B_m\}$ that are disjoint and jointly exhaustive. Each block B_j defines a projection operator $B_j \cap () : \wp(U) \to \wp(U)$ on the powerset $\wp(U)$ as a vector

---

[1] Since the subsets dualize to partitions, the Yoga correlates subspaces to direct-sum decompositions and thus the Birkhoff-von-Neumann quantum logic of subspaces (Birkhoff and Von Neumann 1936) has a correlated quantum logic of DSDs (Ellerman 2018).

Fig. 3.1 A subset S
uniquely represented as
union (sum) of projections
$B_j \cap S$

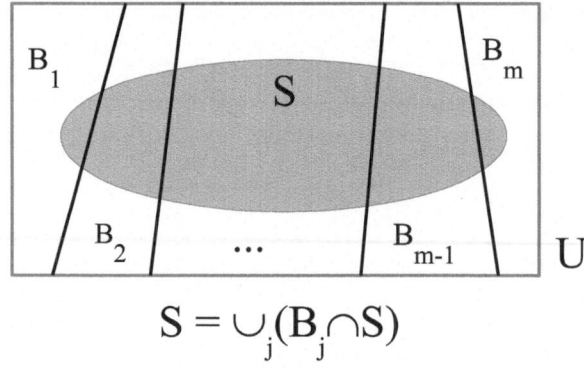

$$S = \cup_j (B_j \cap S)$$

space $\mathbb{Z}_2^n$ over $\mathbb{Z}_2$ (with set addition as the symmetric difference (Ellerman 2017)) down to the subspace $\wp(B_j)$. Then the following statements are equivalent:

1. $\{B_1, B_2, ..., B_m\}$ is a set partition of U;
2. every non-empty subset $S \subseteq U$, is uniquely expressed as the union of subsets of the B_j (i.e., the non-empty projections $B_j \cap S$, $j = 1, ..., m$ as in Fig. 3.1); and
3. $\{\wp(B_j)\}_{j=1}^m$ is a DSD of the vector space $\wp(U) \cong \mathbb{Z}_2^n$.

Thinking of $B_j \cap () : \wp(U) \to \wp(U)$ as a projection operator, the set-level 'resolution of the identity' is:

$$I() = \cup_j \left[B_j \cap () \right] : \wp(U) \to \wp(U).$$

This running of the Yoga backwards raises the question of what is the set-analogue of the notion of an eigenvector and eigenvalue. For $r \in \mathbb{R}$ and $S \subseteq U$, let "rS" stand for "the value r is assigned to the elements of S". Then we have the eigenvector/eigenvalue equation for $f : U \to \mathbb{R}$: (where $f \upharpoonright S$ is the restriction of f to S)

$$f \upharpoonright S = rS \text{ in analogy with } Fv = rv.$$

Thus the set-notion of an eigenvector is just a constant set and the set notion of an eigenvalue is that constant value on a constant set as illustrated in Table 3.3.

A characteristic function $\chi_S : U \to 2 \subseteq \mathbb{R}$, defined for a subset S of the basis set U, defines a projection operator $P_{[S]} : V \to V$ to the subspace $[S]$ generated by $S \subseteq U$ defined by $P_{[S]} u_i = \chi_S(u_i) u_i$ for $u_i \in U$, where the eigenvalue function for the operator $P_{[S]}$ is the characteristic function $\chi_S : U \to 2 = \{0, 1\}$ for S. The constant sets for the characteristic function $\chi_S : U \to \mathbb{R}$ are $S = \chi_S^{-1}(1)$ and $S^c = U - S = \chi_S^{-1}(0)$, and the corresponding eigenvalues of $P_{[S]}$ are 1 and 0 for the eigenspaces $[S]$ and $[S^c]$. If $P_{V_r} : V \to V$ is the projection operator to an eigenspace V_r for a diagonalizable operator $F : V \to V$, then the spectral decomposition of F is $\sum_r r P_{V_r}$ that is prefigured by the 'spectral decomposition' of the numerical attribute f:

Table 3.3 Set-level concepts and the corresponding Hilbert space concepts

Set concept	Hilbert-space concept
Partition $\left\{ f^{-1}(r) \right\}_{r \in f(U)}$	DSD $\{V_r\}_{r \in f(U)}$
$U = \uplus_{r \in f(U)} f^{-1}(r)$	$V = \oplus_{r \in f(U)} V_r$
Numerical attribute $f : U \to \mathbb{R}$	Observable $F u_i = f(u_i) u_i$
$f \upharpoonright S = rS$	$F u_i = r u_i$
Constant set S of f	Eigenvector u_i of F
Value r on constant set S	Eigenvalue r of eigenvector u_i
Characteristic fcn. $\chi_S : U \to \{0, 1\}$	Proj. operator $P_{[S]} u_i = \chi_S(u_i) u_i$
$\cup_{r \in f(U)} \left(f^{-1}(r) \cap () \right) = I : \wp(U) \to \wp(U)$	$\sum_{r \in f(U)} P_{V_r} = I : V \to V$
Spectral Decomp. $f = \sum_{r \in f(U)} r \chi_{f^{-1}(r)}$	Spectral Decomp. $F = \sum_r r P_{V_r}$
Set of r-constant sets $\wp\left(f^{-1}(r) \right)$	Eigenspace V_r of r-eigenvectors
Partition join $\pi \vee \sigma$	Lüders op. $\hat{\rho} = \sum_{r \in f(U)} P_{V_r} \rho P_{V_r}$

Table 3.4 Direct product of sets corresponds to the tensor product of vector spaces

Set concept	Vector-space concept
Direct product $U \times U'$	Tensor product $V \otimes W$
(u_i, u'_k) element of $U \times U'$	Basis element $u_i \otimes u'_k$ of $V \otimes W$

$$f = \sum_{r \in f(U)} r \chi_{f^{-1}(r)} : U \to \mathbb{R}.$$

The Yoga of Linearization starts with a set concept and then, applied to a basis set, it generates the corresponding vector space concept by taking all linear combinations of the basis vectors using scalars from the base field. Starting with an observable $F : V \to V$ defined on an orthonormal (ON) basis set U of eigenvectors, we can go backwards to arrive at the numerical attribute $f : U \to \mathbb{R}$ as the eigenvalue function assigning the eigenvalue to each eigenvector in the basis set.

Partitions underlie both quantum states and quantum observables. When we characterized a non-singleton block of a partition as the skeletal version of a superposition state, that was in the context of the partition treatment of quantum states. The blocks in the inverse-image partition $\left\{ f^{-1}(r) \right\}_{r \in f(U)}$ of $f : U \to \mathbb{R}$ do not prefigure quantum states but prefigure the subsets of eigenvectors having the same eigenvalue where the cardinality of the block represents the dimension of the eigenspace. The following Table 3.3 gives the partition-math/Hilbert-space math translation dictionary.

What is the vector space version of the Cartesian or direct product of sets $U \times U'$? Our Yoga is: apply the set concept to basis sets and see what it generates. Let U be a basis for V and U' be a basis for W, then the set product of basis sets is $U \times U'$ and they (bi)linearly generate the tensor product $V \otimes W$ with the ordered pair (u_i, u'_k) elements of $U \times U'$ corresponding to the basis elements $u_i \otimes u'_k$ of $V \otimes W$ as given in Table 3.4.

3.3 The Intermediate Yoga: From Sets to Vectors-as-Sets in QM/Sets

Since the Yoga can be applied using vector spaces over any field, we could take the field to be $\mathbb{Z}_2 = \{0, 1\}\}$ to give an intermediate stage between the set concepts and the vector space over $\mathbb{C}$ concepts, namely vector spaces over $\mathbb{Z}_2$. The integers mod 2, $\mathbb{Z}_2$, form a field and $\mathbb{Z}_2^n$ is the n-dimensional vector space of n-ary 0, 1-column vectors over $\mathbb{Z}_2$. For any n-element set $U = \{u_1, ..., u_n\}$, there is an isomorphism $\mathbb{Z}_2^n \cong \wp(U)$ where the standard basis vectors of $\mathbb{Z}_2^n$ are mapped to the singletons of powerset $\wp(U)$. The non-singleton subsets in the powerset $\wp(U)$ are to be interpreted as superposition subsets (previously analyzed), not subsets of distinct elements. In $\wp(U)$ as a vector space over $\mathbb{Z}_2$ for $S, T \in \wp(U)$, the addition of subsets is the symmetric difference (or inequivalence) $S \Delta T$ so the vector sum in terms of subsets is:

$$S + T = S\Delta T = (S - T) \cup (T - S) = \left(S \cap T^c\right) \cup \left(T \cap S^c\right) = S \cup T - (S \cap T).$$

This means that the addition operation on subsets just deletes the common elements–which corresponds to the addition mod 2 of the 0, 1-column vectors corresponding to S and T. The column vector corresponding to a subset $S \subseteq U$ is $[\chi_S(u_1), ..., \chi_S(u_n)]^t$ (where the superscript indicates the transpose). The cardinality $|S\Delta T|$ is the *Hamming distance* in coding theory, the number of places where the vectors differ, which could also be expressed as the Euclidean distance (squared), $|S\Delta T| = \sum_{u_i \in U} (\chi_S(u_i) - \chi_T(u_i))^2$, that prefigures the distance between vectors in QM/$\mathbb{C}$.

The Yoga of Linearization carries the set-based mathematics of partitions to the vector-space mathematics of quantum mechanics which uses inner product spaces over $\mathbb{C}$. But the intermediate stage of carrying the set-based mathematics of partitions to vector spaces over $\mathbb{Z}_2$ gives the added benefit of developing a pedagogical or 'toy' model of QM over $\mathbb{Z}_2$, i.e., quantum mechanics over sets (QM/sets) (Ellerman 2017). Hence some aspects of QM/sets will be developed here and in later chapters. QM/$\mathbb{Z}_2$ or QM/sets is the vector space version of the skeletal model at the set level since $\mathbb{Z}_2^n \cong \wp(U)$ is the vector space where the vectors are sets. In short:

QM/sets = Skeletal set-level model of QM as a vector space over $\mathbb{Z}_2$.

We can take $U = \{a, b, c\}$ and $\{\{a\}, \{b\}, \{c\}\}$ as the computational basis (so $\{a\} \leftrightarrow [1, 0, 0]^t$, etc.), but there are many other bases for $\wp(U)$, all of which could be expressed in the computational basis. For instance, $\{\{a, b\}, \{a, b, c\}, \{b, c\}\}$ is a basis since:

$\{a, b, c\} + \{b, c\} = \{a\}$;
$\{a, b\} + \{a, b, c\} + \{b, c\} = \{b\}$; and
$\{a, b\} + \{a, b, c\} = \{c\}$.

Then taking $\left\{a'\right\} = \{a, b\}$, $\left\{b'\right\} = \{a, b, c\}$, and $\left\{c'\right\} = \{b, c\}$, we have the U'-basis for $U' = \left\{a', b', c'\right\}$. Since a "ket" in QM is an abstract vector independent of its

Table 3.5 Ket table giving a vector space isomorphism: $\mathbb{Z}_2^3 \cong \wp(U) \cong \wp(U') \cong \wp(U'')$ where row = ket

$\mathbb{Z}_2^3$	$U = \{a, b, c\}$	$U' = \{a', b', c'\}$	$U'' = \{a'', b'', c''\}$
$[1, 1, 1]^t$	$\{a, b, c\}$	$\{b'\}$	$\{a'', b'', c''\}$
$[1, 1, 0]^t$	$\{a, b\}$	$\{a'\}$	$\{b''\}$
$[0, 1, 1]^t$	$\{b, c\}$	$\{c'\}$	$\{b'', c''\}$
$[1, 0, 1]^t$	$\{a, c\}$	$\{b', c'\}$	$\{c''\}$
$[1, 0, 0]^t$	$\{a\}$	$\{b', c'\}$	$\{a''\}$
$[0, 1, 0]^t$	$\{b\}$	$\{a', b', c'\}$	$\{a'', b''\}$
$[0, 0, 1]^t$	$\{c\}$	$\{a', b'\}$	$\{a'', c''\}$
$[0, 0, 0]^t$	$\emptyset$	$\emptyset$	$\emptyset$

representation in a basis, we can illustrate that notion with a "ket table" showing the same ket in different bases across a row–as in Table 3.5.

Gauss's formula $(2^n - 1)(2^n - 2)(2^n - 2^2) \ldots (2^n - 2^{n-1})$ (Lidl and Niederreiter 1986, p. 71) is for the number of ordered bases in $\mathbb{Z}_2^n$ and the number of different unordered bases is obtained by dividing by $n!$. In the case of $n = 3$, $(2^3 - 1)(2^3 - 2)(2^3 - 4) = 7 \times 6 \times 4$ ordered bases of $\mathbb{Z}_2^3$ so dividing by $3! = 6$ gives $7 \times 4 = 28$ different bases for $\mathbb{Z}_2^3$, three of which are listed in Table 3.5.

In ordinary QM, i.e., QM/$\mathbb{C}$, the bracket $\langle \psi | \phi \rangle$ is defined as the (basis independent) inner product, but there are no inner products in vector spaces over finite fields. However, the interpretation of $\langle \psi | \phi \rangle$ is the overlap between the states, and the natural notion of overlap in $\wp(U)$ is the (basis dependent) intersection $\langle S |_U T \rangle = |S \cap T|$, i.e., the cardinality of the intersection of the two sets expressed in the indicated basis. The same overlap is given by the scalar product of the column vectors for S and T, i.e., $\langle S |_U T \rangle = \sum_{u_i \in U} \chi_S(u_i) \chi_T(u_i)$ with the sum computed in the natural numbers. The ket $|T\rangle$ is basis independent (Table 3.5), but the bra $\langle S |_U$ is basis-dependent (since the bracket $\langle S |_U T \rangle$ requires the two kets $|S\rangle$ and $|T\rangle$ to be expressed in the same basis to form the intersection $S \cap T$).

The ket-bra combination $|\{u\}\rangle \langle \{u\} |_U$ for $u \in U$ applied to the ket $|T\rangle$ has the value $|\{u\}\rangle \langle \{u\} |_U T \rangle = |\{u_i\}\rangle$ if $u_i \in T$ and is the zero vector $\emptyset$ otherwise, so it is the projection operator to the subspace $\{\emptyset, \{u_i\}\}$ which in set terms is: $\{u_i\} \cap () : \wp(U) \to \wp(U)$ where $\{u_i\} \cap T$ is either $\{u_i\}$ or $\emptyset$. The sum $\sum_{u_i \in S} |\{u_i\}\rangle \langle \{u_i\} |_U$ is the projection operator $S \cap () : \wp(U) \to \wp(U)$, and the sum $\sum_{u_i \in U} |\{u_i\}\rangle \langle \{u_i\} |_U = I_U : \wp(U) \to \wp(U)$, is the identity operator–all prefiguring the corresponding machinery in QM/$\mathbb{C}$.

The setting of $\mathbb{Z}_2^n \cong \wp(U)$ allows us to make an interesting separation between DSDs and the eigenspaces of (always diagonalizable) linear operators. For any numerical attribute $f : U \to \mathbb{R}$, there is an inverse image partition on U which induces subspaces $\{\wp(f^{-1}(r))\}_{r \in f(U)}$ that form a direct-sum decomposition of $\wp(U)$. But a linear operator on $\mathbb{Z}_2^n$ has to have eigenvalues in the base field so the only linear operators are the projection operators $S \cap () : \wp(U) \to \wp(U)$ which have only binary DSDs with the subspaces generated by S and S^c. The general role of observables in QM/sets is taken by real-valued numerical attributes on U and their DSDs can have many eigenspaces–hinting that the basic notion of observable is the DSD of eigenspaces (Ellerman 2018), not the linear operator generating it in QM/$\mathbb{C}$ (see the later treatment of commuting operators which bears the same message).

For the resolution of unity, $\langle S|_U \{u_i\}\rangle = \langle\{u_i\}|_U S\rangle = |\{u_i\} \cap S| = \chi_S(u_i)$, so

$$\langle S|_U T\rangle = \sum_{u_i \in U} \langle S|_U \{u_i\}\rangle \langle\{u_i\}|_U T\rangle = \sum_{u_i \in U} \chi_S(u_i)\,\chi_T(u_i) = |S \cap T|$$

which is the QM/sets version of the usual $\langle\psi|\varphi\rangle = \sum_i \langle\psi|u_i\rangle\langle u_i|\varphi\rangle$ in QM/$\mathbb{C}$.

The usual norm $|\psi| = \sqrt{\langle\psi|\psi\rangle}$ notation conflicts with our use of $|S|$ for cardinality so we will use the notation $\|S\|_U := \sqrt{\langle S|_U S\rangle} = \sqrt{|S|}$ for the (basis-dependent) norm in QM/sets. In QM/$\mathbb{C}$, vectors can be normalized at any time, but in QM/sets, the normalization awaits the computation of probabilities so we will use the corresponding formulas in ordinary QM for unnormalized $|\psi\rangle$. When a state $|\psi\rangle$ is measured in the measurement basis $\{|u_i\rangle\}_{i=1}^n$, then the Born Rule gives:

$$\Pr(|u_i\rangle \mid |\psi\rangle) = \frac{\|\langle u_i|\psi\rangle\|^2}{\|\psi\|^2}$$

so the corresponding Born rule in QM/sets is:

$$\Pr(\{u_i\}|_U S) = \frac{\|\langle\{u_i\}|_U S\rangle\|_U^2}{\|S\|_U^2} = \frac{|\{u_i\} \cap S|}{|S|} = \begin{cases} 1/|S| & \text{if } u_i \in S \\ 0 & \text{if } u_i \notin S \end{cases}.$$

This is the usual conditional probability for the outcome $\{u_i\}$ in an equiprobable outcome space U conditioned on the event S. But in QM/sets, it is interpreted as the probability of the outcome $\{u_i\}$ when measuring the superposition state S represented by the pure state density matrix $\rho(S)$ with the entries $\rho(S)_{ik} = \frac{1}{|S|}$ if $u_i, u_k \in S$ and 0 otherwise, i.e., $\Pr(\{u_i\}|S) = \text{tr}\left[P_{\{u_i\}}\rho(S)\right]$.

If we have a numerical attribute $f : U \to \mathbb{R}$ (e.g., a real-valued random variable on an outcome set U), then:

$$\|S\|_U^2 = \langle S|_U S\rangle = \sum_{r \in f(U)} \left\|f^{-1}(r) \cap S\right\|_U^2 = \sum_{r \in f(U)} \left|f^{-1}(r) \cap S\right| = |S|.$$

Then the probability of getting the value r in a drawing from the outcome subset or event S in QM/sets is:

$$\Pr(r|_U S) = \frac{\left\|f^{-1}(r) \cap S\right\|_U^2}{\|S\|_U^2} = \frac{\left|f^{-1}(r) \cap S\right|}{|S|}$$

Table 3.6 Intermediate Yoga of Linearization

Set concepts	Vectors-as-sets concepts
Universe set U	Basis $\{\{u_i\}\}_{u_i \in U}$ for $\wp(U)$
Cardinality $|U|$	Dimension of $\wp(U)$
Subset $S \subseteq U$	Subspace $[S] = \wp(S) \cong 2^S$ of $\wp(U)$
Char. fcn. $\chi_S : U \to \mathbb{Z}_2$	Proj. op. $S \cap () : \wp(U) \to \wp(U)$
$\left\{f^{-1}(r)\right\}_{r \in f(U)}$ of $f : U \to \mathbb{R}$	DSD $\left\{\wp\left(f^{-1}(r)\right)\right\}_{r \in f(U)}$ of $\wp(U)$
Value r in $f(U)$	Eigenvalue r associated with $\wp\left(f^{-1}(r)\right)$
Constant set $S \subseteq f^{-1}(r)$	Eigenset S in $\wp\left(f^{-1}(r)\right)$
$S, T \subseteq U$, overlap $|S \cap T|$	Scalar prod. $\sum_{u_i} \chi_S(u_i) \chi_T(u_i) = \langle S|_U T \rangle$
Norm $\|S\| = \sqrt{S}$, $\|S\|^2 = |S|$	$\|S\|_U = \sqrt{\chi_S \cdot \chi_S}$, $\|S\|_U^2 = |S|$
Hamming dist. $|S \Delta T|$	$|S \Delta T| = \sum_{u_i \in U} (\chi_S(u_i) - \chi_T(u_i))^2$
Direct prod. $U \times U'$	Tensor prod. $\wp(U \times U') \cong \wp(U) \otimes \wp(U')$
Elements $(u_i, u'_k) \in U \times U'$	Basis $\{u_i\} \otimes \{u'_k\}$ for $\wp(U) \otimes \wp(U')$

which corresponds to the usual formula for an eigenvalue r:

$$\Pr(r|\psi) = \frac{\left\|P_{V_r}(\psi)\right\|^2}{\|\psi\|^2}$$

where $P_{V_r}(\psi)$ is the projection of ψ to the eigenspace of r just as $f^{-1}(r) \cap S$ is the projection of S to the eigenspace determined by r, namely the subspace $\wp\left(f^{-1}(r)\right)$ of $\wp(U)$. It could also be obtained by: $\Pr(r|S) = \text{tr}\left[P_{f^{-1}(r)}\rho(S)\right]$ where $P_{f^{-1}(r)}$ is the diagonal projection matrix with the diagonal elements $\chi_{f^{-1}(r)}(u_i)$.

Given two sets U and U' of cardinality n and n', the Cartesian or direct product $U \times U'$ is the set of ordered pairs (u_i, u'_k) for $u_i \in U$ and $u'_k \in U'$. The two vector spaces $\mathbb{Z}_2^n \cong \wp(U)$ and $\mathbb{Z}_2^{n'} \cong \wp(U')$ have a tensor product, and it is just $\wp(U \times U') \cong \wp(U) \otimes \wp(U')$ where the basis element $\{(u_i, u'_k)\} \in \wp(U \times U')$ corresponds to the basis element $\{u_i\} \otimes \{u'_k\}$ of $\wp(U) \otimes \wp(U')$.

The Table 3.6 gives a summary of the intermediate Yoga that linearizes set concepts to vectors-as-sets concepts in $\mathbb{Z}_2^n \cong \wp(U)$.

Then the 'other half' of the Yoga can be presented as going to vector-as-sets to vector-space concepts in an inner product space over $\mathbb{k}$–as in Table 3.7.

Table 3.7 The remaining Yoga

Vectors-as-sets concepts	Vector-space concepts
Basis $\{\{u_i\}\}_{u_i \in U}$ for $\wp(U)$	Basis $\{u_i\}_{u_i \in U}$ for $[U] = V$ over $\Bbbk$
Dimension of $\wp(U)$	Dimension of V
Subspace $[S] = \wp(S) \cong 2^S$ of $\wp(U)$	Subspace $[S] \cong \Bbbk^S$ of V
Proj. op. $S \cap () : \wp(U) \to \wp(U)$	Proj. op. $P_{[S]} : V \to V$
DSD $\left\{\wp\left(f^{-1}(r)\right)\right\}_{r \in f(U)}$ of $\wp(U)$, $f : U \to \mathbb{R}$	DSD $\left\{\left[f^{-1}(r)\right]\right\}_{r \in f(U)}$
Eigenvalue r assoc. with $\wp\left(f^{-1}(r)\right)$	Eigenvalue r of $F : V \to V$
Eigenset S in $\wp\left(f^{-1}(r)\right)$	Eigenvector v of F in $\left[f^{-1}(r)\right]$
Scalar prod. $\sum_{u_i} \chi_S(u_i) \chi_T(u_i) = \langle S \mid_U T \rangle$	Inner product $\langle v \mid v' \rangle$
Norm $\|S\|_U = \sqrt{\chi_S \cdot \chi_S}$, $\|S\|_U^2 = |S|$	Norm $\|v\| = \sqrt{\langle v \mid v \rangle}$
Hamming $|S \Delta T| = \sum_{u_i \in U} (\chi_S(u_i) - \chi_T(u_i))^2$	$\|v - v'\|^2 = \sum_i (v_i - v_i')(v_i - v_i')^*$
Tensor prod. $\wp(U \times U') \cong \wp(U) \otimes \wp(U')$	Tensor prod. $[U] \otimes [U'] = V \otimes V'$
Basis $\{u_i\} \otimes \{u_k'\}$ for $\wp(U) \otimes \wp(U')$	Basis $u_i \otimes u_k'$ for $V \otimes V'$

3.4   The Double-Slit Experiment in QM/Sets

Richard Feynman's favorite experiment to bring out the weirdness of quantum mechanics was the double-slit experiment. But the basic points of the experiment would soon be lost on students if the full mathematics of that experiment was developed, so simplified models are used in the literature. The point of a pedagogical model of QM, such as QM/sets, is of course not to reproduce the full model, but to give in a systematic context, a simplified model that makes the key points.

For instance in the double-slit experiment, if no distinction is made between the particle going through one slit or the other (i.e., no detectors at the slits), then the two states in the superposition are schematically represented by:

$$|\text{going through slit 1}\rangle + |\text{going through slit 2}\rangle$$

evolve unitarily and will show interference effects. The mathematics of the unitary (i.e., no distinctions) evolution over the complex numbers will have a complex-valued "wave interpretation" (Stone's Theorem) without there being any physical waves; the interference results from the addition of vectors representing the parts of the evolving superposition.

When no distinctions are involved, the evolving indefinite superposition does not reach the level of definiteness of the particle going through one slit or the other–which conflicts with our 'classical' intuitive always-definite view of the particle's trajectory ("It has to go through one slit or the other!"). A better intuitive picture would be to recognize the *levels of indefiniteness* at the set level in the lattice of partitions–which would prefigure levels of indefiniteness at the quantum level.

For instance in the simplest model of the particle at the double-slit screen, there are the three definite states:

Fig. 3.2 Schematic diagram
for the double-slit
experiment with detection at
the slits

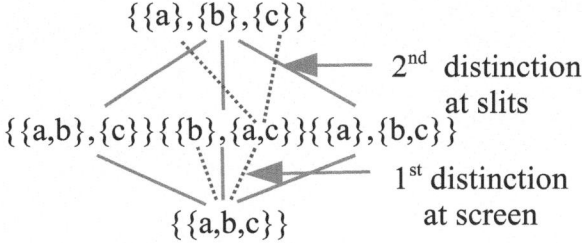

$$\{a\} = |\text{going through slit 1}\rangle,$$

$\{a\} = |\text{going through slit 1}\rangle,$

$\{b\} = |\text{hitting the screen between slits}\rangle,$ and

$\{c\} = |\text{going through slit 2}\rangle.$

Consider the schematic partition lattice on these three states.[2] Before arriving at the screen containing the two slits, the particle is in the superposition state $\{\{a, b, c\}\}$. The first level of distinction, i.e., the first measurement, distinguishes between the particle hitting the screen at $\{b\}$ or remaining in the superposition $\{a, c\}$ consisting of a and c. This first distinction or 'measurement' at the screen takes the pure state $\{\{a, b, c\}\}$ to the mixed state $\{\{b\}, \{a, c\}\}$. If the particle does not hit the screen, the state reduction from the mixed state $\{\{b\}, \{a, c\}\}$ is to the state $\{a, c\}$. If a further distinction or measurement is made at the slits, then the particle has to go through slit 1 (state $\{a\}$) or through slit 2 (state $\{c\}$) as illustrated in Fig. 3.2.

Our classical reasoning is "The particle has to go through slit 1 or slit 2 to get to the other side of the screen"; that reasoning assumes fully distinguished states like in the classical mixture $\mathbf{1}_U$ at the top of the lattice. But that reasoning only applies when a distinction is made between the two slits which would determine which slit the particle went through. That measurement pushes the level of distinction up to the classical level $\{\{a\}, \{c\}\}$ and the resulting probability distribution at the wall is obtained classically. But in the absence of any such distinction, the quantum state stays in the superposition state $\{\{a, c\}\}$ and it is the superposition state that unitarily evolves, not the mixed classical state $\{\{a\}, \{c\}\}$. That leads to the interference effects resulting from the non-zero off-diagonal terms in the density matrix for the pure superposition state $\{a, c\}$ as a block in the partition $\{\{b\}, \{a, c\}\}$ (in the equiprobable case since $\frac{1/3}{2/3} = \frac{1}{2}$):

$$\rho(\{a, c\}) = \begin{bmatrix} \frac{1}{2} & 0 & \frac{1}{2} \\ 0 & 0 & 0 \\ \frac{1}{2} & 0 & \frac{1}{2} \end{bmatrix}$$

with non-zero off-diagonal terms indicating the superposition between $\{a\}$ and $\{c\}$.

Our classical fully-definite intuitions only 'see' the world in terms of the top fully-definite discrete partition, as it were, so the unitary evolution of the superposition state, $\{a, c\}$, i.e., $|\text{going through slit1}\rangle + |\text{going through slit2}\rangle$, is hard to imagine as a

---

[2] A "shorthand" has been used to keep some figures where $|U| = 4$ manageable, but we will avoid it in the treatment of the double-slit experiment where $|U| = 3$.

Fig. 3.3 Incorrect intuition that particle has to go through one slit or the other before evolution (dashed arrows) versus seeing evolution (solid arrow) at the level of the superposition

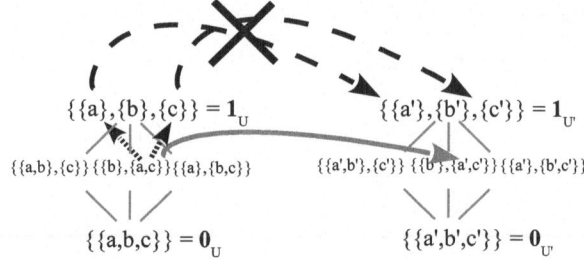

classical real-world process–as if quantum evolution could only take place at the level of full distinctions and definiteness ("It has to go through one slit or the other!")–as illustrated in Fig. 3.3.

In the classical world, everything has to happen in terms of fully distinguished states (no superposition states), so the interference effects of a single particle evolving from the superposition state seems a 'mystery.' Yes, in the classical world of fully distinguished states (i.e., only discrete partitions as it were), the particle has to go through one slit or the other to get to the detection wall, but quantum mechanics tells us that there are superposition states (at a lower level of indefiniteness in the lattice picture) that evolve unitarily to reach the detection wall and give the quantum interference effects.

To begin to imagine quantum evolution, we have to stop forcing all imagery of change to the fully distinct classical level; the skeletal partition lattice model provides a schematic alternative. Like in the metaphor of 3D printing, a quantum state is built 'from below'. To understand the double-slit experiment, we have to imagine evolution at the lower level of distinction; it is the superposition state $\{a, c\}$ (lower down in the lattice) that evolves (when no distinction is made at the slits), not the classical mixed state $\{\{a\}, \{c\}\}$ as shown in Fig. 3.3. Since the evolution took place at the level of the objectively indefinite superposition $\{a, c\}$, there was no classical-level fact-of-the-matter of the particle going through slit a or slit c.[3]

Anthony Leggett has a good way to bring out the key point. When speaking to a physics-department audience, he considers the double-slit experiment in the case where there is no detection at slit 1 or 2 (or path A or B). An ensemble of identically prepared atoms is fired at the slits, one by one, so that the striped interference pattern slowly develops. Then he asks for a vote on the negative statement "that it is not the case that each individual atom of the relevant ensemble chooses either path A or path B." And he reports that "I almost invariably get a large majority in favor." (Leggett 2011, pp. 154–5). Hence it would seem that most physicists realize that if each atom went through one slit of the other, then the striped interference pattern would not appear. The explanation given here, illustrated in Fig. 3.3, explains the type of non-classical evolution that yields the striped interference pattern (see below) without resort to the classical picture of the atom having to go through one slit or the other.

---

[3] This is unlike the analysis in Bohmian mechanics which strives to flatter our classical intuitions that the particle has to go through one slit or the other.

Fig. 3.4 Measurement at the gates (owl must go through one gate or the other)

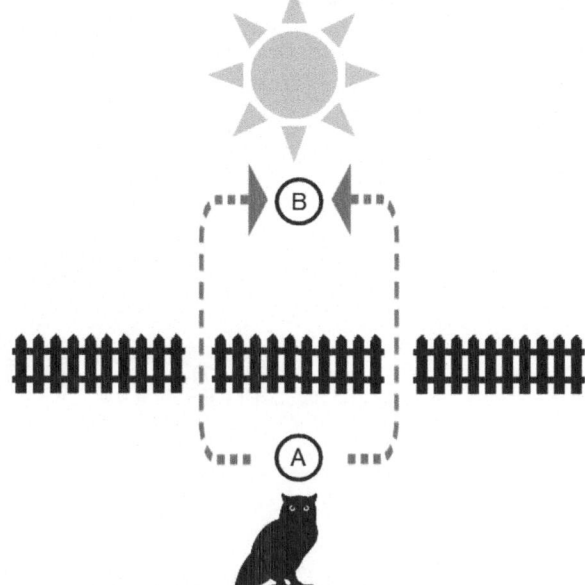

Measurement at gates

Perhaps a metaphor will help to imagine this evolution at a non-classical level. Consider Hegel's Owl of Minerva (who only flies at night) (Hegel 1967, p. 13) facing a high fence with two gates as in the double slit experiment. When the sun is shining, that is a metaphor for detection at the gates to see which gate the Owl of Minerva must walk through to get from A to B in Fig. 3.4 since the owl is then limited to horizontal 'classical' trajectory on the "flatland" (Abbott and Stewart 2008) (the flatland metaphor is also used by Kastner (2015) to make similar points).

But with no light source or sun (i.e., no detection), then the Owl of Minerva has an indefinite flight trajectory and can go from A to B without going through a gate as in Fig. 3.5.

Our classical ("flatlander") intuitions see only the definite ground-level paths or trajectories and, in the absence of the projecting light source (detection at the gates), will still insist on asking: "Which gate did the owl go through?". But with no detections at the slits in the double-slit experiment, there is no matter of fact of the particle going through a slit at the classical level since the evolution is at the non-classical quantum level (which allows interference) as in Fig. 3.3 (illustrated by the third dimension in our flatlander metaphor of Fig. 3.5). This quantum evolution of one superposition $\{a, c\}$ into another $\{a', c'\}$ cannot be understood as long as one insists on interpreting QM in terms of our biologically evolved intuitions based on the macroscopic (superposition-less) world.

In QM/$\mathbb{C}$, the unitary dynamics can be characterized as taking an orthonormal (ON) basis to an ON basis or as simply preserving the values of the brackets (or inner

Fig. 3.5 No measurement at
gates (owl can fly in third
dimension from A to B)

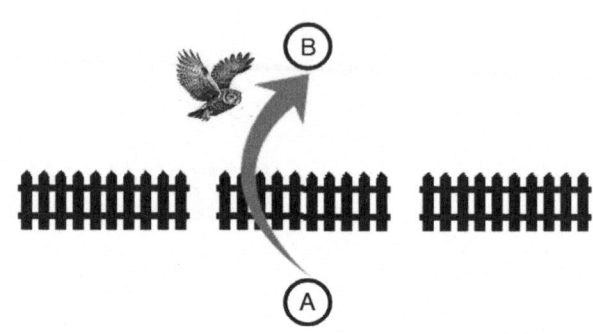

products) $\langle\psi|\phi\rangle$. The analogue of taking an ON basis to an ON basis in QM/sets
is just taking a basis set to a basis set, i.e., a non-singular linear transformation
$\mathbb{Z}_2^n \to \mathbb{Z}_2^n$.

In the transformation from the U-basis $\{\{a\},\{b\},\{c\}\}$ to the U'-basis
$\{\{a'\},\{b'\},\{c'\}\}$, the non-singular linear transformation $\{a\} \longmapsto \{a'\}, \{b\} \longmapsto$
$\{b'\}, \{c\} \longmapsto \{c'\}$ would preserve the size of intersections each expressed in its
own basis. In that manner, non-singular transformation also preserve the brackets,
e.g., $\langle\{a\}\,|_U\,\{a,b\}\rangle = \langle\{a'\}\,|_{U'}\,\{a',b'\}\rangle$, just as unitary transformations preserve the
Dirac brackets in QM/$\mathbb{C}$. This is how QM/Sets mimics the probability calculations
of QM/$\mathbb{C}$. Hence non-singular transformations will be taken as the 'dynamics' in
QM/sets (which was also done in modal QM (Schumacher and Westmoreland 2012))
for each discrete time period.

We take a non-singular dynamics that most closely mimics the spreading of waves
as in the wavey math of QM/$\mathbb{C}$. For $U = \{a,b,c\}$, consider the dynamics: $\{a\} \to$
$\{a'\} = \{a,b\}$; $\{b\} \to \{b'\} = \{a,b,c\}$; and $\{c\} \to \{c'\} = \{b,c\}$ in one time period.
This is represented by the non-singular one-period change of state matrix:

$$A = C_{U' \leftarrow U} = \begin{bmatrix} 1 & 1 & 0 \\ 1 & 1 & 1 \\ 0 & 1 & 1 \end{bmatrix}.$$

To illustrate the double slit experiment, we again take a, b, and c as three vertical
positions and we have a vertical screen with slits at a and c as in Fig. 3.6. Then there
is a wall to the right of the slits so that evolution from the double-slit barrier to the
wall in one time period is according to the A-dynamics.

First case of distinctions at slits: The first case is where we measure the U-state
at the slits and then let the resultant position eigenstate evolve by the A-dynamics

Fig. 3.6 Setup for two-slit
experiment

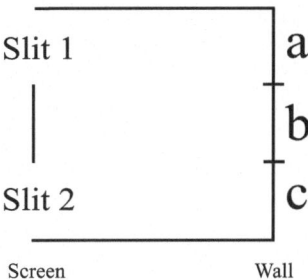

to hit the wall at the right where the vertical position is measured again. Prior to the measurement at the slits, the particle is in a superposition state $\{a, c\}$ between slit 1 at $\{a\}$ and slit 2 at $\{c\}$. The probability that the particle is detected at slit 1 or at slit 2 is:

$$\Pr\left(\{a\} \text{ at slits} \mid \{a, c\} \text{ at slits}\right) = \frac{\|\langle\{a\} |_U \{a, c\}\rangle\|_U^2}{\|\{a, c\}\|_U^2} = \frac{|\{a\} \cap \{a, c\}|}{|\{a, c\}|} = \frac{1}{2};$$

$$\Pr\left(\{c\} \text{ at slits} \mid \{a, c\} \text{ at slits}\right) = \frac{\|\langle\{c\} |_U \{a, c\}\rangle\|_U^2}{\|\{a, c\}\|_U^2} = \frac{|\{c\} \cap \{a, c\}|}{|\{a, c\}|} = \frac{1}{2}.$$

If the particle was at slit 1, i.e., was in eigenstate $\{a\}$, then it evolves in one time period by the A-dynamics to $\{a, b\}$ where the position measurements yield the probabilities of being at a or at b as:

$$\Pr\left(\{a\} \text{ at wall} \mid \{a, b\} \text{ at wall}\right) = \frac{\|\langle\{a\} |_U \{a, b\}\rangle\|_U^2}{\|\{a, b\}\|_U^2} = \frac{|\{a\} \cap \{a, b\}|}{|\{a, b\}|} = \frac{1}{2}$$

$$\Pr\left(\{b\} \text{ at wall} \mid \{a, b\} \text{ at wall}\right) = \frac{\|\langle\{b\} |_U \{a, b\}\rangle\|_U^2}{\|\{a, b\}\|_U^2} = \frac{|\{b\} \cap \{a, b\}|}{|\{a, b\}|} = \frac{1}{2}.$$

If the particle was found in the first measurement to be at slit 2, i.e., was in eigenstate $\{c\}$, then it evolved in one time period by the A-dynamics to $\{b, c\}$ where the position measurements yield the probabilities of being at b or at c as:

$$\Pr\left(\{b\} \text{ at wall} \mid \{b, c\} \text{ at wall}\right) = \frac{|\{b\} \cap \{b, c\}|}{|\{b, c\}|} = \frac{1}{2}$$

$$\Pr\left(\{c\} \text{ at wall} \mid \{b, c\} \text{ at wall}\right) = \frac{|\{c\} \cap \{b, c\}|}{|\{b, c\}|} = \frac{1}{2}.$$

Since there were distinctions made at both slits, then we add *probabilities* by the Feynman rule (distinguishable case (Feynman 1951)) to compute the probabilities of the particle being measured at the three positions on the wall at the right (see

Fig. 3.7 Particle distribution
at the wall with detection at
slits

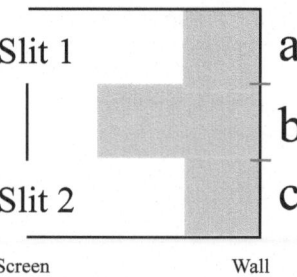

Fig. 3.8 Particle distribution
with no detection at slits

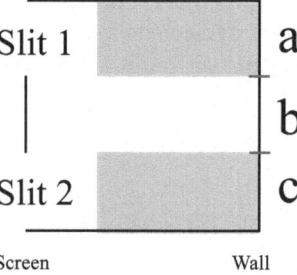

Fig. 3.7) if it starts at the slits in the superposition state $\{a, c\}$ *and* the measurements
were made at the slits:

$$\Pr(\{a\} \text{ at wall} \mid \{a, c\} \text{ at slits}) = \tfrac{1}{2} \times \tfrac{1}{2} = \tfrac{1}{4};$$
$$\Pr(\{b\} \text{ at wall} \mid \{a, c\} \text{ at slits}) = \tfrac{1}{2} \times \tfrac{1}{2} + \tfrac{1}{2} \times \tfrac{1}{2} = \tfrac{1}{2};$$
$$\Pr(\{c\} \text{ at wall} \mid \{a, c\} \text{ at slits}) = \tfrac{1}{2} \times \tfrac{1}{2} = \tfrac{1}{4}.$$

This probability distribution at the detection wall is like the one resulting from adding
the probabilities resulting from a classical particle going through slit 1 with one-half
probability to the probabilities resulting from the particle going through slit 2 with
one-half probability.

   Second case of no distinctions at slits: The second case is when no measurements
are made at the slits and then the superposition state $\{a, c\}$ evolves linearly by the
A-dynamics to $\{a, b\} + \{b, c\} = \{a, c\}$ where the adding of amplitudes at $\{b\}$ cancels
out. Then the final probabilities will just be probabilities of finding $\{a\}$, $\{b\}$, or $\{c\}$
when the measurement is made only at the wall on the right are pictured in Fig. 3.8:

$$\Pr(\{a\} \text{ at wall} \mid \{a, c\} \text{ at slits}) = \Pr(\{a\} \mid \{a, c\}) = \tfrac{|\{a\} \cap \{a, c\}|}{|\{a, c\}|} = \tfrac{1}{2};$$
$$\Pr(\{b\} \text{ at wall} \mid \{a, c\} \text{ at slits}) = \Pr(\{b\} \mid \{a, c\}) = \tfrac{|\{b\} \cap \{a, c\}|}{|\{a, c\}|} = 0;$$
$$\Pr(\{c\} \text{ at wall} \mid \{a, c\} \text{ at slits}) = \Pr(\{c\} \mid \{a, c\}) = \tfrac{|\{c\} \cap \{a, c\}|}{|\{a, c\}|} = \tfrac{1}{2}.$$

   The particle distribution shows the usual stripes of the double-slit experiment in
QM/$\mathbb{C}$ for the no-distinctions-at-the-slits case. In that second case, no state reduction

took place at the slits due to no distinctions being made there, so the indistinct element $\{a, c\}$ evolved (rather than one or the other of the distinct elements $\{a\}$ or $\{c\}$). The action of A is the same on $\{a\}$ and $\{c\}$ as when they evolve separately since A is a linear operator but the two *amplitudes* are now added together *as part of the evolution* (instead of the probabilities adding as in the first case–see the later discussion of the Feynman rules). This allows the interference of the two results and thus the cancellation of the $\{b\}$ term in $\{a\} + \{c\} \overset{A}{\leadsto} \{a, b\} + \{b, c\} = \{a, c\}$. The addition is, of course, mod 2 (where $-1 = +1$) so, in "wave language," the "wave crest" and "wave trough" that add at the location $\{b\}$ cancel out. But these vectors in $\mathbb{Z}_2^3 \cong \wp(U)$ are not waves, and yet the key points in the double-slit experiment come out clearly in the pedagogical model of QM/sets. This reinforces our previous remarks that the most misleading imagery for QM is the wave imagery of the wave interpretation–which is an artifact of the formulation in a vector space over $\mathbb{C}$ (the natural mathematics to describe waves). Again, the point is not that the wave mathematics is wrong since a vector space over $\mathbb{C}$ will always have a 'wave' interpretation (i.e., amplitude + phase in the polar representation), but that the wave imagery should be replaced with the indefiniteness imagery of density matrices prefigured in the mathematics of partitions–where distinctions and indistinctions are key concepts.

3.5 Entanglement in QM/Sets

Starting with a vector-space concept in QM/$\mathbb{C}$, we consider the corresponding set concept which is then restated in $\mathbb{Z}_2^n$ where the vectors are interpreted as subsets–and thus QM/Sets is generated. In QM/$\mathbb{C}$, the notion of entanglement arises by considering the tensor product of two single particle quantum systems–each represented by a state in a vector space. In QM/$\mathbb{C}$, entanglement provides examples of non-local and holistic properties of quantum systems.

What is the set-version of the tensor product of vector spaces? For sets, the notion of product is the Cartesian or direct product of two sets: $X \times Y = \{(x, y) : x \in X, y \in Y\}$. Applying the Yoga, we take the direct product of a basis set $\{v_1, ..., v_m\}$ of a V and a basis set $\{w_1, ..., w_m\}$ of W (over the same base field). Instead of representing each ordered pair as (v_i, w_j), we denote that it as $v_i \otimes w_j$ and then we bilinearly generate a vector space from those basis pairs. That vector space is the tensor product $V \otimes W$. Thus the tensor product of vector spaces is the vector-space version of the direct product of sets–even though there is also the notion of the direct product of vector spaces $V \times W$. The vector space direct product $V \times W$ is just the set of ordered pairs $\{(v, w) : v \in V, w \in W\}$ with component-wise addition and it has a basis set of $\{(v_1, 0), (v_2, 0), ..., (v_m, 0)\} \cup \{(0, w_1), (0, w_2), ..., (0, w_n)\}$ so it has dimension $\dim(V) + \dim(W) = m + n$. In contrast, the tensor product $V \otimes W$ has dimension $|\{v_1, ..., v_m\} \times \{w_1, ..., w_m\}| = \dim(V) \times \dim(W) = mn$.

We have seen that the direct or Cartesian product of sets lifts by the Yoga to the tensor product of vector spaces. When the vector spaces are over $\mathbb{Z}_2$, then the set concept and the vector-space concept are 'essentially' the same. That is, given two finite sets $X = \{x_1, ..., x_m\}$ and $Y = \{y_1, ..., y_n\}$ the cardinality of $X \times Y$ is $|X| \times |Y| = mn$ and the dimension of $\wp(X) \otimes_{\mathbb{Z}_2} \wp(Y)$ is $\dim(\wp(X)) \times \dim(\wp(Y)) = mn$ where the basis elements $x_i \otimes y_j$ of the tensor product are correlated with the ordered pairs $(x_i, y_i) \in X \times Y$ in the vector space isomorphism:

$$\wp(X) \otimes_{\mathbb{Z}_2} \wp(Y) \cong \wp(X \times Y)$$

as vector spaces over $\mathbb{Z}_2$. Hence for the sake of simplicity, we will just use the direct product $X \times Y$ as the universe set to analyze entanglement in QM/Sets.

In QM/$\mathbb{C}$, a vector $z \in V \otimes W$ is said to be *separated* if there are vectors $v \in V$ and $w \in W$ such that $z = v \otimes w$; otherwise, z is said to be *entangled*. In QM/Sets, a subset $S \subseteq X \times Y$ is said to be *separated* or a *product set* if there exists subsets $S_X \subseteq X$ and $S_Y \subseteq Y$ such that $S = S_X \times S_Y$; otherwise $S \subseteq X \times Y$ is said to be *entangled*. In general, let S_X be the support or projection of S on X, i.e., $S_X = \{x : \exists y \in Y, (x, y) \in S\}$ and similarly for S_Y. Then S is separated iff $S = S_X \times S_Y$.

For any subset $S \subseteq X \times Y$, a natural measure of its "entanglement" can be constructed by first viewing S as the support of the equiprobable or Laplacian joint probability distribution on S. If $|S| = N$, then define $\Pr(x, y) = \frac{1}{N}$ if $(x, y) \in S$ and $\Pr(x, y) = 0$ otherwise.

The marginal distributions[4] are defined in the usual way:

$$\Pr(x) = \sum_y \Pr(x, y)$$

$$\Pr(y) = \sum_x \Pr(x, y).$$

A joint probability distribution $\Pr(x, y)$ on $X \times Y$ is *independent* if for all $(x, y) \in X \times Y$,

$$\Pr(x, y) = \Pr(x)\Pr(y).$$

Independent distribution

Otherwise $\Pr(x, y)$ is said to be *correlated*–in which case $\Pr(x, y)$ might be said to have 'holistic' properties that cannot be reconstructed from $\Pr(x)$ and $\Pr(y)$.

Proposition 3.1 *A subset $S \in \wp(X \times Y)$ is entangled iff the equiprobable distribution on S is correlated (non-independent).*

Proof If S is separated, i.e., $S = S_X \times S_Y$, then $\Pr(x) = |S_Y|/N$ for $x \in S_X$ and $\Pr(y) = |S_X|/N$ for $y \in S_Y$ where $|S_X||S_Y| = |S| = N$. Then for $(x, y) \in S$,

$$\Pr(x, y) = \frac{1}{N} = \frac{N}{N^2} = \frac{|S_X||S_Y|}{N^2} = \Pr(x)\Pr(y)$$

---

[4] The marginal distributions are the set versions of the reduced density matrices of QM/$\mathbb{C}$.

Table 3.8 The six entangled subsets of $\wp\left(U \times U\right)$ and corresponding vectors in $\wp\left(U\right) \otimes_{\mathbb{Z}_2} \wp\left(U\right)$

$S \in \wp\left(U \times U\right)$	$v \in \wp\left(U\right) \otimes \wp\left(U\right)$
$\{(a,a),(b,b)\}$	$\{a\} \otimes \{a\} + \{b\} \otimes \{b\}$
$\{(a,b),(b,a)\}$	$\{a\} \otimes \{b\} + \{b\} \otimes \{a\}$
$\{(a,a),(a,b),(b,a)\}$	$\{a\} \otimes \{a\} + \{a\} \otimes \{b\} + \{b\} \otimes \{a\}$
$\{(a,a),(a,b),(b,b)\}$	$\{a\} \otimes \{a\} + \{a\} \otimes \{b\} + \{b\} \otimes \{b\}$
$\{(a,b),(b,a),(b,b)\}$	$\{a\} \otimes \{b\} + \{b\} \otimes \{a\} + \{b\} \otimes \{b\}$
$\{(a,a),(b,a),(b,b)\}$	$\{a\} \otimes \{a\} + \{b\} \otimes \{a\} + \{b\} \otimes \{b\}$

and $\Pr(x,y) = 0 = \Pr\left(x\right)\Pr\left(y\right)$ for $(x,y) \notin S$ so the equiprobable distribution is independent. If S is entangled, i.e., $S \neq S_X \times S_Y$, then $S \subsetneq S_X \times S_Y$ so let $(x,y) \in S_X \times S_Y - S$. Then $\Pr\left(x\right), \Pr\left(y\right) > 0$ but $\Pr\left(x,y\right) = 0$ so it is not independent, i.e., is correlated. □

Consider the set version of one qubit space where $U = \{a,b\}$. The product set $U \times U$ has four elements and thus $\wp\left(U \times U\right)$ has 15 nonempty subsets. Each $\wp\left(U\right)$ has 3 nonempty subsets so $3 \times 3 = 9$ of the 15 subsets are "separated" subsets leaving 6 "entangled" subsets listed in Table 3.8.

The first two might be called *Bell states* which are the two graphs of bijections $U \longleftrightarrow U$. The degree of entanglement can be measured by the logical divergence between $\Pr\left(u,u'\right)$ and $\Pr\left(u\right)\Pr\left(u'\right)$ for $u, u' \in U$ where for two probability distributions $p = \{p_1, ..., p_n\}$ and $q = \{q_1, ..., q_n\}$, the *logical divergence* between them is defined as the square of the Euclidean distance between them: $d\left(p\|q\right) = \sum_i \left(p_i - q_i\right)^2$ (Ellerman 2021, p. 31).

Consider the Bell state $S = \{(a,b),(b,a)\}$. Then $\Pr\left(u,u'\right) = \frac{1}{|S|} = \frac{1}{2}$ if $\left(u,u'\right) \in S$ and $\Pr\left(u,u'\right) = 0$ for $u,u' \in U$ otherwise so $\Pr(a,b) = \frac{1}{2} = \Pr(b,a)$ while $\Pr\left(u,u'\right) = 0$ otherwise. Since $X = Y = U$, we need to, in general, distinguish the marginals on the left and right. In this symmetrical case, the marginals on the left and right are: $\Pr_L(a) = \sum_{u \in U} \Pr(a,u) = \frac{1}{2} = \sum_{u \in U} \Pr(u,a) = \Pr_R(a)$ and $\Pr(b)_L = \sum_{u \in U} \Pr(b,u) = \frac{1}{2} = \sum_{u \in U} \Pr(u,b) = \Pr_R(b)$. Then the logical divergence of the Bell state is:

$$d(\Pr(-,-) \| \Pr_L(-)\Pr_R(-))$$
$$= \sum_{(u,u') \in U \times U} \left(\Pr\left(u,u'\right) - \Pr(u)_L \Pr_R\left(u'\right)\right)^2$$
$$= \left(\Pr(a,a) - \Pr(a)_L \Pr(a)_R\right)^2 + \left(\Pr(a,b) - \Pr(a)_L \Pr(b)_R\right)^2 +$$
$$\left(\Pr(b,a) - \Pr(b)_L \Pr(a)_R\right)^2 + \left(\Pr(b,b) - \Pr(b)_L \Pr(b)_R\right)^2$$
$$= \left(-\tfrac{1}{2}\tfrac{1}{2}\right)^2 + \left(\tfrac{1}{2} - \tfrac{1}{2}\tfrac{1}{2}\right)^2 + \left(\tfrac{1}{2} - \tfrac{1}{2}\tfrac{1}{2}\right)^2 + \left(-\tfrac{1}{2}\tfrac{1}{2}\right)^2$$
$$= \tfrac{1}{16} + \tfrac{1}{16} + \tfrac{1}{16} + \tfrac{1}{16} = \tfrac{1}{4}.$$

Consider the entangled state $S = \{(a,a),(a,b),(b,a)\}$ which is not a Bell state. Then $\Pr\left(u,u'\right) = \frac{1}{|S|} = \frac{1}{3}$ if $\left(u,u'\right) \in S$ and 0 otherwise. The marginals on the left are $\Pr_L(a) = \sum_{u \in U} \Pr(a,u) = \frac{2}{3}$ and $\Pr_L(b) = \sum_{u \in U} \Pr(b,u) = \frac{1}{3}$ and the marginals on the right are: $\Pr_R(a) = \sum_{u \in U} \Pr(u, a) = \frac{2}{3}$ and $\Pr_R(b) = \sum_{u \in U} \Pr(u, b) = \frac{1}{3}$. Then the logical divergence is:

$$d(\Pr(-, -) \| \Pr_L(-) \Pr_R(-)) = \sum_{(u, u') \in U \times U} \left(\Pr(u, u') - \Pr_L(u) \Pr_R(u')\right)^2$$
$$= \left(\frac{1}{3} - \frac{2}{3}\frac{2}{3}\right)^2 + \left(\frac{1}{3} - \frac{2}{3}\frac{1}{3}\right)^2 + \left(\frac{1}{3} - \frac{1}{3}\frac{2}{3}\right)^2 + \left(-\frac{1}{3}\frac{1}{3}\right)^2$$
$$= \frac{1}{81} + \frac{1}{81} + \frac{1}{81} + \frac{1}{81} = \frac{4}{81}.$$

And for one of the separated states, say, $S = \{a\} \times \{a, b\} = \{(a, a), (a, b)\}$, we have $\Pr(a, a) = \Pr(a, b) = \frac{1}{2}$ and 0 otherwise. $\Pr_L(a) = \sum_{u \in U} \Pr(a, u) = 1$ and $\Pr_R(a) = \frac{1}{2}$ while $\Pr_L(b) = 0$ and $\Pr_R(b) = \frac{1}{2}$. Hence the logical divergence is:

$$d(\Pr(-, -) \| \Pr_L(-) \Pr_R(-)) = \sum_{(u, u') \in U \times U} \left(\Pr(u, u') - \Pr_L(u) \Pr_R(u')\right)^2$$
$$= \left(\frac{1}{2} - \frac{1}{1}\frac{1}{2}\right)^2 + \left(\frac{1}{2} - \frac{1}{1}\frac{1}{2}\right)^2 + \left(0 - 0\frac{1}{2}\right)^2 + \left(0 - 0\frac{1}{2}\right)^2 = 0.$$

This logical divergence $d(\Pr(-, -) \| \Pr_L(-) \Pr_R(-))$ as a measure of entanglement gives the maximal entanglement $\frac{1}{4}$ to the Bell states, an intermediate entanglement $\frac{4}{81}$ to the entangled states that are not Bell states, and zero entanglement to the separated states.

For an entangled subset S, a sampling u of left-hand system will change the probability distribution for a sampling of the right-hand system u', $\Pr(u'|u) \neq \Pr(u')$. In the case of maximal entanglement (e.g., the Bell states), when S is the graph of a bijection between U and U, e.g., $\{(a, b), (b, a)\}$ is the graph of the bijection $U \cong U$ where $a \longleftrightarrow b$), the value of the right-hand u' is determined by the value of left-hand u (and vice-versa). The Bell state $\{(a, b), (b, a)\}$ in $\wp(U \times U)$ or $\{a\} \otimes \{b\} + \{b\} \otimes \{a\}$ in $\wp(U) \otimes_{\mathbb{Z}_2} \wp(U)$ is the QM/Sets version of the entangled Bell state $|L \downarrow\rangle \otimes |R \uparrow\rangle + |L \uparrow\rangle \otimes |R \downarrow\rangle$ in the standard Bell-type experiment in QM/$\mathbb{C}$ where two particles are separated, one going left and the other going right. When one measures the spin of the left-hand particle, then it determines the spin of the right-hand particle.

3.6 Bell's Theorem in QM/Sets

This treatment of Bell's Theorem in QM/Sets is based on the simple expository example developed by Bernard D 'Espagnat (1979). A simple version of a Bell inequality can be derived in the case of $\mathbb{Z}_2^2$ with three bases $U = \{a, b\}, U' = \{a', b'\}$, and $U'' = \{a'', b''\}$, and where the kets are listed in Table 3.9.

The different basis vectors can be thought of as spin-up and spin-down along three different A, B, and C axes, e.g. $\{a\} = \{A^+\}$ and $\{b\} = \{A^-\}$, $\{a'\} = \{B^+\}$ and $\{b'\} = \{B^-\}$, and $\{a''\} = \{C^+\}$ and $\{b''\} = \{C^-\}$, but we will stick to our usual notation.

Attributes or observables defined on the three universe sets U, U', and U'', such as say $\chi_{\{a\}}$, $\chi_{\{b'\}}$, and $\chi_{\{a''\}}$, are incompatible as can be seen in several ways. For

Table 3.9 Ket table for $\wp\left(U\right) \cong \wp\left(U'\right) \cong \wp\left(U''\right) \cong \mathbb{Z}_2^2$

Kets	U-basis	U'-basis	U''-basis
$|1\rangle$	$\{a,b\}$	$\{a'\}$	$\{a''\}$
$|2\rangle$	$\{b\}$	$\{b'\}$	$\{a'',b''\}$
$|3\rangle$	$\{a\}$	$\{a',b'\}$	$\{b''\}$
$|4\rangle$	$\emptyset$	$\emptyset$	$\emptyset$

instance the set partitions defined on U and U', namely $\{\{a\},\{b\}\}$ and $\{\{a'\},\{b'\}\}$, cannot be obtained as two different ways to partition the same set since $\{a\} = \{a',b'\}$ and $\{a'\} = \{a,b\}$, i.e., an eigenstate in one basis is a superposition in the other. The same holds in the other pairwise comparison of U and U'' and of U' and U''.

A more technical way to show incompatibility is to exploit the vector space structure of $\mathbb{Z}_2^2$ and to see if the projection matrices for $\{a\} \cap ()$ and $\{b'\} \cap ()$ commute. The basis conversion matrices between the U-basis and U'-basis are:

$$\mathcal{C}_{U \leftarrow U'} = \begin{bmatrix} 1 & 0 \\ 1 & 1 \end{bmatrix} \text{ and } \mathcal{C}_{U' \leftarrow U} = \begin{bmatrix} 1 & 0 \\ 1 & 1 \end{bmatrix}.$$

The projection matrix for $\{a\} \cap ()$ in the U-basis is, of course, $\begin{bmatrix} 1 & 0 \\ 0 & 0 \end{bmatrix}$ and the projection matrix for $\{b'\} \cap ()$ in the U'-basis is $\begin{bmatrix} 0 & 0 \\ 0 & 1 \end{bmatrix}$. Converting the latter to the U-basis to check commutativity gives:

$$\mathcal{C}_{U \leftarrow U'} \begin{bmatrix} 0 & 0 \\ 0 & 1 \end{bmatrix} \mathcal{C}_{U' \leftarrow U}$$
$$= \begin{bmatrix} 1 & 0 \\ 1 & 1 \end{bmatrix} \begin{bmatrix} 0 & 0 \\ 0 & 1 \end{bmatrix} \begin{bmatrix} 1 & 0 \\ 1 & 1 \end{bmatrix} = \begin{bmatrix} 0 & 0 \\ 1 & 1 \end{bmatrix}.$$

Hence the commutativity check is:

$$\begin{bmatrix} 1 & 0 \\ 0 & 0 \end{bmatrix} \begin{bmatrix} 0 & 0 \\ 1 & 1 \end{bmatrix} = \begin{bmatrix} 0 & 0 \\ 0 & 0 \end{bmatrix} \neq$$
$$\begin{bmatrix} 0 & 0 \\ 1 & 1 \end{bmatrix} \begin{bmatrix} 1 & 0 \\ 0 & 0 \end{bmatrix} = \begin{bmatrix} 0 & 0 \\ 1 & 0 \end{bmatrix}$$

so the two operators for the observables $\chi_{\{a\}}$ and $\chi_{\{b'\}}$ do not commute. In a similar manner, it is seen that the three observables are mutually incompatible.

Given a ket in $\mathbb{Z}_2^2 \cong \wp\left(U\right) \cong \wp\left(U'\right) \cong \wp\left(U''\right)$, and using the equiprobability assumption on different basis sets, the probabilities of getting the different outcomes for the various observables in the different given states are given in the Table 3.10.

Table 3.10 State-outcome table

State\Outcome of test	$\chi_{\{a\}} = 1$	$= 0$	$\chi_{\{b'\}} = 0$	$= 1$	$\chi_{\{a''\}} = 1$	$= 0$
$\{a, b\} = \{a'\} = \{a''\}$	$\frac{1}{2}$	$\frac{1}{2}$	1	0	1	0
$\{b\} = \{b'\} = \{a'', b''\}$	0	1	0	1	$\frac{1}{2}$	$\frac{1}{2}$
$\{a\} = \{a', b'\} = \{b''\}$	1	0	$\frac{1}{2}$	$\frac{1}{2}$	0	1

Since $\wp(U) \otimes \wp(U) \cong \wp(U \times U)$, the vectors in the tensor product are the subsets of direct product of sets (as seen in the above treatment of entanglement in QM/Sets). Thus in the U-basis, the basis elements are the elements of $U \times U$ and the vectors are all the subsets in $\wp(U \times U)$. But we could obtain the same space as $\wp(U' \times U')$ and $\wp(U'' \times U'')$, and we can construct a ket table where each row is a ket expressed in the different bases. And these calculations in terms of sets could also be carried out in terms of vector spaces over $\mathbb{Z}_2$ where the rows of the ket table are the kets in the tensor product:

$$\mathbb{Z}_2^2 \otimes \mathbb{Z}_2^2 \cong \wp(U \times U) \cong \wp(U' \times U') \cong \wp(U'' \times U'').$$

Since $\{a\} = \{a', b'\} = \{b''\}$ and $\{b\} = \{b'\} = \{a'', b''\}$, the subset $\{a\} \times \{b\} = \{(a, b)\} \in \wp(U \times U)$ is expressed in the $U' \times U'$-basis as $\{a', b'\} \times \{b'\} = \{(a', b'), (b', b')\}$, and in the $U'' \times U''$-basis it is $\{b''\} \times \{a'', b''\} = \{(b'', a''), (b'', b'')\}$. Hence one row in the ket table has:

$$\{(a, b)\} = \{(a', b'), (b', b')\} = \{(b'', a''), (b'', b'')\}.$$

Since the full ket table has 16 rows, we will just give a partial Table 3.11 that suffices for our calculations.

As before, we can classify each vector or subset as separated or entangled and we can furthermore see how that is independent of the basis. For instance $\{(a, a), (a, b)\}$ is separated since:

Table 3.11 Partial ket table for $\wp(U \times U) \cong \wp(U' \times U') \cong \wp(U'' \times U'')$

$\wp(U \times U)$	$\wp(U' \times U')$	$\wp(U'' \times U'')$
$\{(a, a)\}$	$\{(a', a'), (a', b'), (b', a'), (b', b')\}$	$\{(b'', b'')\}$
$\{(a, b)\}$	$\{(a', b'), (b', b')\}$	$\{(b'', a''), (b'', b'')\}$
$\{(b, a)\}$	$\{(b', a'), (b', b')\}$	$\{(a'', b''), (b'', b'')\}$
$\{(b, b)\}$	$\{(b', b')\}$	$\{(a'', a''), (a'', b''), (b'', a''), (b'', b'')\}$
$\{(a, a), (a, b)\}$	$\{(a', a'), (b', a')\}$	$\{(b'', a'')\}$
$\{(a, a), (b, a)\}$	$\{(a', a'), (a', b')\}$	$\{(a'', b'')\}$
$\{(a, a), (b, b)\}$	$\{(a', a'), (a', b'), (b', a')\}$	$\{(a'', a''), (a'', b''), (b'', a'')\}$
$\{(a, b), (b, a)\}$	$\{(a', b'), (b', a')\}$	$\{(a'', b''), (b'', a'')\}$

$$\{(a, a), (a, b)\} = \{a\} \times \{a, b\} = \{(a', a'), (b', a')\} = \{a', b'\} \times \{a'\} =$$
$$\{(b'', a'')\} = \{b''\} \times \{a''\}.$$

An example of an entangled state is:

$$\{(a, a), (b, b)\} = \{(a', a'), (a', b'), (b', a')\} = \{(a'', a''), (a'', b''), (b'', a'')\}.$$

Taking this entangled state as the initial state, there is a probability distribution on $U \times U' \times U''$ where $\Pr(a, a', a'')$ (for instance) is defined as the probability of getting the result $\{a\}$ if a U-measurement is performed on the left-hand system, and if instead a U'-measurement is performed on the left-hand system then $\{a'\}$ is obtained, and if instead a U''-measurement is performed on the left-hand system then $\{a''\}$ is obtained. Thus we would have $\Pr(a, a', a'') = \frac{1}{2} \times \frac{2}{3} \times \frac{2}{3} = \frac{2}{9}$. In this way the probability distribution $\Pr(x, y, z)$ is defined on $U \times U' \times U''$.

A Bell inequality can be obtained from this joint probability distribution over the outcomes $U \times U' \times U''$ of measuring these three incompatible attributes (D'Espagnat 1979). Consider the following marginals:

$$\Pr(a, a') = \Pr(a, a', a'') + \Pr(a, a', b'') \checkmark$$
$$\Pr(b', b'') = \Pr(a, b', b'') \checkmark + \Pr(b, b', b'')$$
$$\Pr(a, b'') = \Pr(a, a', b'') \checkmark + \Pr(a, b', b'') \checkmark.$$

The two terms in the last marginal are each contained in one of the two previous marginals (as indicated by the check marks) and all the probabilities are non-negative, so we have the following inequality:

$$\Pr(a, a') + \Pr(b', b'') \geq \Pr(a, b'')$$
Bell inequality.

All this has to do with measurements on the left-hand system. But there is an alternative interpretation to the probabilities $\Pr(x, y)$, $\Pr(y, z)$, and $\Pr(x, z)$ *if* we assume that the outcome of a measurement on the right-hand system is *independent* of the outcome of the same measurement on the left-hand system. Then $\Pr(a, a')$ is the probability of a U-measurement on the left-hand system giving $\{a\}$ and then a U'-measurement on the right-hand system giving $\{a'\}$, and so forth. Under that *independence assumption* and for this initially prepared Bell state (which is left-right symmetrical in each basis),

$$\{(a, a), (b, b)\} = \{(a', a'), (a', b'), (b', a')\} = \{(a'', a''), (a'', b''), (b'', a'')\},$$

the probabilities would be the same.[5] That is, under that assumption, the probabilities, $\Pr(a) = \frac{1}{2} = \Pr(b)$, $\Pr(a') = \frac{2}{3} = \Pr(a'')$, and $\Pr(b') = \frac{1}{3} = \Pr(b'')$ are the same regardless of whether we are measuring the left-hand or right-hand system of that composite state. Thus with those left-right independent measurements, $\Pr(a, a') = \frac{1}{2} \times \frac{2}{3} = \frac{1}{3}$ which is now interpreted as the probability of getting $\{a\}$ in a left-hand U-measurement and then getting $\{a'\}$ in a right-hand U'-measurement, and similarly $\Pr(b', b'') = \frac{1}{3} \times \frac{1}{3} = \frac{1}{9}$, and $\Pr(a, b'') = \frac{1}{2} \times \frac{1}{3} = \frac{1}{6}$, so the above Bell inequality would still hold. But we can use QM/$\mathbb{Z}_2$ to compute the probabilities for those different measurements on the two systems to see if the independence assumption is compatible with QM/$\mathbb{Z}_2$.

To compute $\Pr(a, a')$, we first measure the left-hand component in the U-basis. Since $\{(a, a), (b, b)\}$ is the given state, and (a, a) and (b, b) are equiprobable, the probability of getting $\{a\}$ (i.e., the eigenvalue 1 for the observable $\chi_{\{a\}}$) is $\frac{1}{2}$. But the right-hand system is then in the state $\{a\}$ and the probability of getting $\{a'\}$ (i.e., eigenvalue 0 for the observable $\chi_{\{b'\}}$) is $\frac{1}{2}$ (as seen in the state-outcome Table 3.10). Thus the probability is $\Pr(a, a') = \frac{1}{2} \times \frac{1}{2} = \frac{1}{4}$.

To compute $\Pr(b', b'')$, we first perform a U'-basis measurement on the left-hand component of the given state $\{(a, a), (b, b)\} = \{(a', a'), (a', b'), (b', a')\}$, and we see that the probability of getting $\{b'\}$ is $\frac{1}{3}$. Then the right-hand system is in the state $\{a'\}$ and the probability of getting $\{b''\}$ in a U''-basis measurement of the right-hand system in the state $\{a'\}$ is 0 (as seen from the state-outcome Table 3.10). Hence the probability is $\Pr(b', b'') = 0$.

Finally we compute $\Pr(a, b'')$ by first making a U-measurement on the left-hand component of the given state $\{(a, a), (b, b)\}$ and get the result $\{a\}$ with probability $\frac{1}{2}$. Then the state of the second system is $\{a\}$ so a U''-measurement will give the $\{b''\}$ result with probability 1 so the probability is $\Pr(a, b'') = \frac{1}{2}$.

Then we plug the probabilities into the Bell inequality:

$$\Pr(a, a') + \Pr(b', b'') \geq \Pr(a, b'')$$
$$\tfrac{1}{4} + 0 \ngeq \tfrac{1}{2}$$

Violation of Bell inequality.

The violation of the Bell inequality shows that the independence assumption about the measurement outcomes on the left-hand and right-hand systems is incompatible with QM/$\mathbb{Z}_2$ so the effects of the QM/$\mathbb{Z}_2$ measurements are said to be "nonlocal"–or, more simply put, the outcome of the left-measurement affects the outcomes of the right-measurement of the entangled state.

---

[5] The same holds for the other "Bell state": $\{(a, b), (b, a)\}$.

References

Abbott E, Stewart I (2008) The annotated Flatland: a romance of many dimensions. Basic Books, New York

Auletta G, Wang S-Y (2014) Quantum mechanics for thinkers. Pan Stanford Publishing, Singapore

Birkhoff G, Von Neumann J (1936) The logic of quantum mechanics. Ann Math 37:823–43

D'Espagnat B (1979) The quantum theory and reality. Sci Am 241:158–181

Dubreil P, Dubreil-Jacotin M-L (1939) Théorie algébrique des relations d'équivalence. J de Mathématique 18:63–95

Ellerman D (2017) Quantum mechanics over sets: a pedagogical model with non-commutative finite probability theory as its quantum probability calculus. Synthese 194:4863–4896. https://doi.org/10.1007/s11229-016-1175-0

Ellerman D (2018) The quantum logic of direct-sum decompositions: the dual to the quantum logic of subspaces. Logic J IGPL 26:1–13. https://doi.org/10.1093/jigpal/jzx026

Ellerman D (2021) New foundations for information theory: logical entropy and Shannon entropy. Springer Nature, Cham, Switzerland

Feynman RP (1951) The concept of probability in quantum mechanics. In: Second Berkeley symposium on mathematical statistics and probability. University of California Press, pp 533–541

Finberg D, Mainetti M, Rota G-C (1996) The logic of commuting equivalence relations. In: Ursini A, Agliano P (eds) Logic and algebra. Marcel Dekker, New York, pp 69–96

Hegel GWF (1967) Hegel's philosophy of right. Oxford University Press, New York

Hoffman K, Kunze R (1961) Linear algebra. Prentice-Hall, Englewood Cliffs, NJ

Kastner RE (2015) Understanding our unseen reality: solving quantum riddles. Imperial College Press, London

Leggett A (2011) Quantum interview. In: Schlosshauer M (ed) Elegance and enigma: the quantum interviews. Springer, Heidelberg

Lidl R, Niederreiter H (1986) Introduction to finite fields and their applications. Cambridge University Press, Cambridge, UK

Ore O (1942) Theory of equivalence relations. Duke Math J 9:573–627

Rota G-C (1997) Indiscrete thoughts. Birkhauser, Boston

Schumacher B, Westmoreland M (2012) Modal quantum theory. Found Phys 42:918–925

Chapter 4
The Mathematics of Quantum Mechanics: The Partition Analysis

> *The probability of an event (in an ideal experiment where there are no uncertain external disturbances) is the absolute square of a complex quantity called the probability amplitude. When the event can occur in several alternative ways the probability amplitude is the sum of the probability amplitude for each alternative considered separately. If an experiment capable of determining which alternative is actually taken is performed the interference is lost and the probability becomes the sum of the probability for each alternative.*
>
> *Richard Feynman (1951)*

Abstract This chapter gives the partitional treatment of quantum measurement along with a number of other applications such as: commuting and non-commuting observables, von Neumann's two types of quantum processes, the collapse postulate, quantum jumps, Feynman's rules about adding amplitudes or probabilities (and the resulting "state reduction principle"), the partition version of the principle of identity of indistinguishables, Weyl's interesting imagery for measurement, and the indistinguishability of like particles. Then the Yoga is used to systematically extend the concepts of 'classical' logical entropy to the quantum logical entropy which is then shown to naturally measure the results of quantum measurement. Finally the Yoga is again applied to relate group representations on sets (i.e., a group actions) and group representations on vector spaces over $\mathbb{C}$ which turn out to have such important applications in particle physics.

Keywords Non-commuting observables · Conjugate observables · Von Neumann Type I and II processes · Schrödinger equation · Principle of Identity of Indistinguishables · Collapse postulate · Quantum jumps · Indistinguishability of like particles · Quantum logical entropy · Group representations

4.1 Commuting and Non-commuting Observables

4.1.1 Commutativity and Conjugacy for DSDs

Our purpose is to show that partitional mathematics underlies the *mathematics* of quantum mechanics. The emphasis is on the *mathematics*, not the physics of QM. Seeing that the mathematics of QM is the mathematics to describe a reality of objective indefiniteness (not definiteness all the way down) does not provide the physics that must come from the quantization of classical physics.

The one aspect of QM that is often taken as a characteristic of QM to distinguish it from classical physics is non-commuting or even conjugate observables (expressed as Hermitian matrices). At first glance, non-commuting operators seem to have nothing to do with partitions.

The Yoga of Linearization shows that a direct-sum decomposition of a vector space is the vector space version of a partition on a set. Given two Hermitian operators $F, G : V \to V$, let $\{V_i\}_{i \in I}$ be the DSD of eigenspaces for F and let $\{W_j\}_{j \in J}$ be the DSD of eigenspaces for G. We saw in the set case, how the formula for the Lüders mixture operation gives the partition operation of the join, so we may mimic the join operation with the two DSDs in the case of vector spaces. This join-like operation yields the set of non-zero vector spaces $\{V_i \cap W_j\}$ which are the subspaces of the *simultaneous* eigenvectors of F and G.

The join of two partitions on the *same* set yields a partition of that set. For our two Hermitian operators, let $\mathcal{SE}$ be the subspace of V spanned by the non-zero subspaces $\{V_i \cap W_j\}$, i.e., the subspace spanned by the simultaneous eigenvectors of F and G. The point is that $\mathcal{SE}$ need not, in general, be the whole space. The condition specifying whether F and G commute or not is exactly the condition that $\mathcal{SE} = V$ or not. The *commutator* of F and G is: $[F, G] = FG - GF : V \to V$, and as a linear operator on V, the commutator has a kernel ker $[F, G]$ which is the subspace of vectors v such that $[F, G] v = 0$. The operators commute iff $[F, G] = \hat{0}$ (the zero operator) iff ker $([F, G]) = V$.

Proposition 4.1 $\mathcal{SE} = \ker ([F, G])$.

Proof Let $F, G : V \to V$ be two Hermitian operators on a finite dimensional vector space V and let v be a simultaneous eigenvector of the operators, i.e., $Fv = \lambda v$ and $Gv = \mu v$. Then $[F, G] (v) = (FG - GF) (v) = (\lambda\mu - \mu\lambda) v = 0$ so the space $\mathcal{SE}$ spanned by the simultaneous eigenvectors is contained in the kernel ker $([F, G])$, i.e., $\mathcal{SE} \subseteq \ker ([F, G])$. Conversely, if we restrict the two operators to the subspace ker $([F, G])$, then the restricted operators commute on that subspace. Then it is a standard theorem of linear algebra (Hoffman and Kunze 1961, p. 177) that the subspace ker $([F, G])$ is spanned by simultaneous eigenvectors of the two restricted operators. But if a vector is a simultaneous eigenvector for the two operators restricted to a subspace, they are the same for the operators on the whole space V, since the two conditions $Fv = \lambda v$ and $Gv = \mu v$ only involves vectors in the subspace. Hence ker $([F, G]) \subseteq \mathcal{SE}$. $\qquad\square$

Since the condition that the operators commute or not is ker $([F, G]) = V$ or not, it is equivalent to $\mathcal{SE} = V$ or not. But that condition $\mathcal{SE} = V$ deals only with two DSDs with no mention of operators. Thus the commutativity condition on operators is captured by the mathematics of the vector-space version of partitions, i.e., DSDs. And the further condition of the operators being *conjugate* is when $\mathcal{SE} = \mathbf{0}$ (the subspace consisting of only the zero vector). The definitions of commuting, not commuting, and conjugate thus apply to DSDs–even when there are no corresponding operators as, for instance, in vector spaces over $\mathbb{Z}_2$ (see below).

DSDs $\{V_i\}_{i \in I}$ and $\{W_j\}_{j \in J}$ of V are:

$$\textit{commuting if } \mathcal{SE} = V;$$
$$\textit{non-commuting if } \mathcal{SE} \neq V; \text{ and}$$
$$\textit{conjugate if } \mathcal{SE} = \mathbf{0}.$$

The Heisenberg "uncertainty" principle is somewhat mistranslated since "uncertainty" may imply a subjective uncertainty instead of objective indefiniteness. The "principle of indeterminacy" might be a better name.[1] In any case, since conjugate observables have no (non-zero) simultaneous eigenvectors, $\mathcal{SE} = \mathbf{0}$, if a system is in an eigenstate of one observable, it cannot be an eigenstate of a conjugate observable– so the conjugate observable has an indefinite value.

4.1.2 Non-commutativity and Conjugacy in QM/Sets

Since the Yoga shows how the mathematics of indefiniteness can be translated into vector spaces, it might be noted that these properties of commutativity, non-commutativity, and conjugacy are not peculiarly quantum concepts about operators in Hilbert spaces. The concepts can be illustrated in quite simple vector spaces such as $\wp(U) \cong \mathbb{Z}_2^n$.

Example of non-commutativity: Let $U = \{a, b, c\}$ and $U' = \{a', b', c'\}$ where $\{a'\} = \{a, b\}, \{b'\} = \{a, b, c\},$ and $\{c'\} = \{b, c\}$ all as in Table 3.5 with the U-basis being the computational basis. For $f : U \to \mathbb{R}$, let $f(a) = 1$; $f(b) = f(c) = 2$, so the two eigenspaces are $\wp(f^{-1}(1)) = \{\emptyset, \{a\}\}$ and $\wp(f^{-1}(2)) = \{\emptyset, \{b\}, \{c\}, \{b, c\}\}$. For $g : U' \to \mathbb{R}$, lets $g(a') = g(c') = 3$; $g(b') = 5$; so the eigenspaces are $\wp(g^{-1}(3)) = \{\emptyset, \{a'\}, \{c'\}, \{a', c'\}\} = \{\emptyset, \{a, b\}, \{b, c\}, \{a, c\}\}$ and $\wp(g^{-1}(5)) = \{\emptyset, \{b'\}\} = \{\emptyset, \{a, b, c\}\}$. Hence the two direct-sum decompositions of eigenspaces are:

---

[1] Heisenberg's German word was "Unbestimmtheit" which could well be translated as "indefiniteness" or "indeterminacy" rather than "uncertainty".

$$\left\{ \wp\left(f^{-1}(1)\right), \wp\left(f^{-1}(2)\right) \right\}$$
$$= \{\{\emptyset, \{a\}\}, \{\emptyset, \{b\}, \{c\}, \{b, c\}\}\}$$
and
$$\left\{ \wp\left(g^{-1}(3)\right), \wp\left(g^{-1}(5)\right) \right\}$$
$$= \{\{\emptyset, \{a, b\}, \{b, c\}, \{a, c\}\}, \{\emptyset, \{a, b, c\}\}\}.$$

Each attribute has two eigenspaces so there are four intersections and the only non-zero intersection is:

$$\wp\left(f^{-1}(2)\right) \cap \wp\left(g^{-1}(3)\right) = \{\emptyset, \{b\}, \{c\}, \{b, c\}\} \cap \{\emptyset, \{a, b\}, \{b, c\}, \{a, c\}\} = \{\emptyset, \{b, c\}\} = \mathcal{SE}$$

so f and g are non-commuting but are not conjugate.

Example of commutativity: If two numerical attributes f and g are defined on the same set, then of course their DSDs commute. But that is not necessary. Let U and f be the same as above but $U^* = \{a^*, b^*, c^*\}$ where $\{a^*\} = \{a\}$, $\{b^*\} = \{b\}$, and $\{c^*\} = \{b, c\}$ with $g : U^* \to \mathbb{R}$ defined as $g(a^*) = 3$, $g(b^*) = 4$, and $g(c^*) = 5$. Then the three eigenspaces of g (subspaces determined by g) are $\wp\left(g^{-1}(3)\right) = \{\emptyset, a^*\} = \{\emptyset, a\}$, $\wp\left(g^{-1}(4)\right) = \{\emptyset, \{b\}\}$; $\wp\left(g^{-1}(5)\right) = \{\emptyset, \{b, c\}\}$. The direct-sum decompositions are:

$$\left\{ \wp\left(f^{-1}(1)\right), \wp\left(f^{-1}(2)\right) \right\}$$
$$= \{\{\emptyset, \{a\}\}, \{\emptyset, \{b\}, \{c\}, \{b, c\}\}\}$$
and
$$\left\{ \wp\left(g^{-1}(3)\right), \wp\left(g^{-1}(4)\right), \wp\left(g^{-1}(5)\right) \right\}$$
$$= \{\{\emptyset, \{a\}\}, \{\emptyset, \{b\}\}, \{\emptyset, \{b, c\}\}\}.$$

Then the intersections with the eigenspaces of f contain $\{a\}$, $\{b\}$, and $\{b, c\}$ and those three vectors form a basis for $\wp(U) \cong \mathbb{Z}_2^3$ so $\mathcal{SE} = \mathbb{Z}_2^3$ and thus f and g commute.

Example of conjugacy: A simple example of conjugacy can be constructed in $\mathbb{Z}_2^n$ for even numbers $n > 2$. For $n = 4$, consider the U-basis $= \{\{a\}, \{b\}, \{c\}, \{d\}\}$ and $\hat{U}$-basis $= \left\{\{\hat{a}\}, \{\hat{b}\}, \{\hat{c}\}, \{\hat{d}\}\right\} = \{\{b, c, d\}, \{a, c, d\}, \{a, b, d\}, \{a, b, c\}\}$ of $\mathbb{Z}_2^4$ where $\{\hat{a}\} = \{b, c, d\}, ..., \{\hat{d}\} = \{a, b, c\}$. Let $f = \chi_{\{a, b\}} : U \to \mathbb{Z}_2$ so $f(a) = f(b) = 1$ and $f(c) = f(d) = 0$ with the eigenspaces $\wp\left(f^{-1}(1)\right) = \{\emptyset, \{a\}, \{b\}, \{a, b\}\}$ and $\wp\left(f^{-1}(0)\right) = \{\emptyset, \{c\}, \{d\}, \{c, d\}\}$. Let $g = \chi_{\{\hat{b}, \hat{c}\}} : \hat{U} \to \mathbb{Z}_2$ so $g(\hat{b}) = g(\hat{c}) = 1$ and $g(\hat{a}) = g(\hat{d}) = 0$ with the eigenspaces $\wp\left(g^{-1}(1)\right) = \left\{\emptyset, \{\hat{b}\}, \{\hat{c}\}, \{\hat{b}, \hat{c}\}\right\}$ or in the computational basis $\wp\left(g^{-1}(1)\right) = \{\emptyset, \{a, c, d\}, \{a, b, d\}, \{b, c\}\}$ and $\wp\left(g^{-1}(0)\right) = \left\{\emptyset, \{\hat{a}\}, \{\hat{d}\}, \{\hat{a}, \hat{d}\}\right\} = \{\emptyset, \{b, c, d\}, \{a, b, c\}, \{a, d\}\}$. The two direct-sum decompositions are:

$$\left\{ \wp\left(f^{-1}\left(1\right)\right), \wp\left(f^{-1}\left(0\right)\right)\right\}$$
$$= \{\{\emptyset, \{a\}, \{b\}, \{a, b\}\}, \{\emptyset, \{c\}, \{d\}, \{c, d\}\}\}$$

and

$$\left\{ \wp\left(g^{-1}\left(1\right)\right), \wp\left(g^{-1}\left(0\right)\right)\right\}$$
$$= \{\{\emptyset, \{a, c, d\}, \{a, b, d\}, \{b, c\}\}, \{\emptyset, \{b, c, d\}, \{a, b, c\}, \{a, d\}\}\}.$$

Then all the four intersections of eigenspaces have only $\emptyset$ in common so $\mathcal{SE} = \{\emptyset\} = 0$ and f and g are conjugate.

The two numerical attributes in the conjugacy example have been chosen as 0, 1-valued characteristic functions so they define linear operators F and G on $\mathbb{Z}_2^4$ in the usual manner: $Fu_i = f\left(u_i\right)u_i$ and $G\hat{u}_i = g\left(\hat{u}_i\right)\hat{u}_i$. Thus we can compute their commutator as usual once restated in the computational basis. The Fourier-like transformation matrix to convert a 0, 1-vector written in the $\hat{U}$-basis to the same 0, 1-vector written in the U-basis is:

$$C_{U \leftarrow \hat{U}} = \begin{bmatrix} \langle\{a\}\,|_U\,\{\hat{a}\}\rangle & \langle\{a\}\,|_U\,\{\hat{b}\}\rangle & \langle\{a\}\,|_U\,\{\hat{c}\}\rangle & \langle\{a\}\,|_U\,\{\hat{d}\}\rangle \\ \langle\{b\}\,|_U\,\{\hat{a}\}\rangle & \langle\{b\}\,|_U\,\{\hat{b}\}\rangle & \langle\{b\}\,|_U\,\{\hat{c}\}\rangle & \langle\{b\}\,|_U\,\{\hat{d}\}\rangle \\ \langle\{c\}\,|_U\,\{\hat{a}\}\rangle & \langle\{c\}\,|_U\,\{\hat{b}\}\rangle & \langle\{c\}\,|_U\,\{\hat{c}\}\rangle & \langle\{c\}\,|_U\,\{\hat{d}\}\rangle \\ \langle\{d\}\,|_U\,\{\hat{a}\}\rangle & \langle\{d\}\,|_U\,\{\hat{b}\}\rangle & \langle\{d\}\,|_U\,\{\hat{c}\}\rangle & \langle\{d\}\,|_U\,\{\hat{d}\}\rangle \end{bmatrix}$$

$$= \begin{bmatrix} \langle\{a\}\,|_U\,\{b,c,d\}\rangle & \langle\{a\}\,|_U\,\{a,c,d\}\rangle & \langle\{a\}\,|_U\,\{a,b,d\}\rangle & \langle\{a\}\,|_U\,\{a,b,c\}\rangle \\ \langle\{b\}\,|_U\,\{b,c,d\}\rangle & \langle\{b\}\,|_U\,\{a,c,d\}\rangle & \langle\{b\}\,|_U\,\{a,b,d\}\rangle & \langle\{b\}\,|_U\,\{a,b,c\}\rangle \\ \langle\{c\}\,|_U\,\{b,c,d\}\rangle & \langle\{c\}\,|_U\,\{a,c,d\}\rangle & \langle\{c\}\,|_U\,\{a,b,d\}\rangle & \langle\{c\}\,|_U\,\{a,b,c\}\rangle \\ \langle\{d\}\,|_U\,\{b,c,d\}\rangle & \langle\{d\}\,|_U\,\{a,c,d\}\rangle & \langle\{d\}\,|_U\,\{a,b,d\}\rangle & \langle\{d\}\,|_U\,\{a,b,c\}\rangle \end{bmatrix} = \begin{bmatrix} 0&1&1&1 \\ 1&0&1&1 \\ 1&1&0&1 \\ 1&1&1&0 \end{bmatrix}.$$

For instance, $\{a\} = \left\{\hat{b}, \hat{c}, \hat{d}\right\}$ so that vector in $\hat{U}$-basis is the column vector $[0, 1, 1, 1]^t$ and

$$\begin{bmatrix} 0&1&1&1 \\ 1&0&1&1 \\ 1&1&0&1 \\ 1&1&1&0 \end{bmatrix} \begin{bmatrix} 0 \\ 1 \\ 1 \\ 1 \end{bmatrix} \overset{\mathrm{mod}(2)}{=} \begin{bmatrix} 1 \\ 0 \\ 0 \\ 0 \end{bmatrix} = \{a\} \text{ in } U\text{-basis.}$$

The inverse $C_{\hat{U} \leftarrow U}$ to the conversion matrix $C_{U \leftarrow \hat{U}}$ is the same: $C_{\hat{U} \leftarrow U} = C_{U \leftarrow \hat{U}}$; it is orthogonal since the columns are orthogonal and form a basis. Then the projection matrix G to the eigenspace $\wp\left(g^{-1}\left(1\right)\right)$ for the G-operator $\hat{b}, \hat{c} \to 1; \hat{a}, \hat{d} \to 0$ can be converted to the U-basis as:

$$C_{U \leftarrow \hat{U}} G C_{\hat{U} \leftarrow U}$$
$$= \begin{bmatrix} 0&1&1&1 \\ 1&0&1&1 \\ 1&1&0&1 \\ 1&1&1&0 \end{bmatrix} \begin{bmatrix} 0&0&0&0 \\ 0&1&0&0 \\ 0&0&1&0 \\ 0&0&0&0 \end{bmatrix} \begin{bmatrix} 0&1&1&1 \\ 1&0&1&1 \\ 1&1&0&1 \\ 1&1&1&0 \end{bmatrix} \overset{\mathrm{mod}(2)}{=} \begin{bmatrix} 0&1&1&0 \\ 1&1&0&1 \\ 1&0&1&1 \\ 0&1&1&0 \end{bmatrix}.$$

Then we can compute the commutator of F which projects to $\wp\left(f^{-1}(1)\right) = \{\emptyset, \{a\}, \{b\}, (a, b)\}$ and G which projects to $\wp\left(g^{-1}(1)\right) = \{\emptyset, \{a, c, d\}, \{a, b, d\}, \{b, c\}\}$ in the computational U-basis:

$$[F, G] = FG - GF$$

$$= \begin{bmatrix} 1 & 0 & 0 & 0 \\ 0 & 1 & 0 & 0 \\ 0 & 0 & 0 & 0 \\ 0 & 0 & 0 & 0 \end{bmatrix} \begin{bmatrix} 0 & 1 & 1 & 0 \\ 1 & 1 & 0 & 1 \\ 1 & 0 & 1 & 1 \\ 0 & 1 & 1 & 0 \end{bmatrix} - \begin{bmatrix} 0 & 1 & 1 & 0 \\ 1 & 1 & 0 & 1 \\ 1 & 0 & 1 & 1 \\ 0 & 1 & 1 & 0 \end{bmatrix} \begin{bmatrix} 1 & 0 & 0 & 0 \\ 0 & 1 & 0 & 0 \\ 0 & 0 & 0 & 0 \\ 0 & 0 & 0 & 0 \end{bmatrix} \stackrel{\mod(2)}{=} \begin{bmatrix} 0 & 0 & 1 & 0 \\ 0 & 0 & 0 & 1 \\ 1 & 0 & 0 & 0 \\ 0 & 1 & 0 & 0 \end{bmatrix}$$

which has a determinant of 1 so the commutator is a non-singular transformation and thus has a zero space kernel $\mathcal{SE}$.

We illustrated the set-level pure and mixed states using the lattice $\Pi(U)$. The conjugate $\hat{U}$-basis also has a similar lattice $\Pi\left(\hat{U}\right)$ and moving to a more definite state $\{a\} + \{b\} + \{c\} = \{a, b, c\} \rightsquigarrow \{a\}$ in the U-basis would correspond to moving to a less definite state in the conjugate basis $\hat{U}$, e.g., $\{a, b, c\} = \left\{\hat{a}\right\} \rightsquigarrow \left\{\hat{b}, \hat{c}, \hat{d}\right\} = \{a\}$, as one would expect for conjugate bases. This is how the pedagogical model of QM/Sets illustrates the familiar relationship between conjugate observables such as position and momentum in QM.

It might be noted that these numerical attributes with three or more values cannot always be repackaged as linear operators with the eigenvalues in the base field since the only such linear operators on $\mathbb{Z}_2^n$ are projection operators with eigenvalues 0 or 1. But all the concepts of compatibility, i.e., $\mathcal{SE} = V$, incompatibility, i.e., $\mathcal{SE} \neq V$, and conjugacy, i.e., $\mathcal{SE} = \mathbf{0}$, can be defined using the vector-space partitions, i.e., DSDs, in $\mathbb{Z}_2^4$. Thus the mathematics behind "non-commutativity" in QM is not about operators *per se*, but about the underlying vector space *partitions* or DSDs and that is what we have shown.

It is only when $\mathcal{SE} = V$ that the join-like operation taking non-zero intersections of the eigenspaces can properly be called the *join* of the DSDs, otherwise only a *proto-join* when $\mathcal{SE} \neq V$. As Hermann Weyl put it: "Thus combination [DE: join] of two gratings [DE: vector space partitions] presupposes commutability...". (Weyl 1949, p. 257)

Set Version: A set of ordinary set partitions on the same universe U is said to be *complete* if their join is the discrete partition $\mathbf{1}_U = \{\{u_i\}\}_{u_i \in U}$ where the blocks in the join have cardinality one. Numerical attributes defined on the same set are *compatible*, and if they defined a complete set of partitions, they would be a Complete Set of Compatible Attributes (CSCA). If the partitions arose as the inverse-images of numerical attributes, then each element in U would be uniquely characterized by the ordered set of values of the attributes.

For example, consider $f : U = \{a, b, c, d\} \to \mathbb{R}$ where $a, b \longmapsto 1$ and $c, d \longmapsto 2$ as well as $g : U \to \mathbb{R}$ where $a, c \longmapsto 3$ and $b, d \longmapsto 4$. Since $f^{-1} \vee g^{-1} = \{\{a, b\}, \{c, d\}\} \vee \{\{a, c\}, \{b, d\}\} = \{\{a\}, \{b\}, \{c\}, \{d\}\} = \mathbf{1}_U$, those two numerical attributes form a complete set and accordingly, the four elements of U are uniquely

Table 4.1 Set and vector-space versions of commutativity

Set concept	Vector-space concept
Set partition	Direct-sum decomposition (DSD)
Partitions on $U \neq U'$: $|U| = |U'|$	DSDs on V with $\mathcal{SE} \neq V$
Partitions on same set U	DSDs on V with $\mathcal{SE} = V$
Join of partitions on same set U	Join of commuting DSDs
CSCA (blocks have cardinality 1)	CSCO (eigenspaces have dim. 1)

characterized by their ordered pairs of f-values and g-values, i.e., $\{a\}$, $\{b\}$, $\{c\}$, $\{d\}$ are given respectively by: $(1, 3)$, $(1, 4)$, $(2, 3)$, $(2, 4)$.

Quantum Version: Similarly a set of commuting observables is said be *complete* (a CSCO) (Dirac 1958) if all the non-zero intersections of all their eigenspaces, i.e., the subspaces in the join, are of dimension one. Then each of the simultaneous eigenvectors is uniquely characterized by the ordered set of eigenvalues of the operators. With a CSCO, all the distinctions that can be made, have been made. It is not "distinctions all the way down" as in the classical picture of reality. These results are summarized in Table 4.1.

4.2   Quantum Processes and Measurement

4.2.1   Von Neumann's Two Types of Quantum Processes

Quantum concepts need to be 'seen' in a certain way to see the underlying mathematics of indefiniteness and definiteness as shown in the case of non-commuting operators. At first glance, the time-dependent Schrödinger equation to describe the evolution of an isolated system seems to have nothing to do with distinctions.

Von Neumann classified quantum processes into Type 1 (measurement) and Type 2 (evolution described by the Schrödinger equation). We have seen that a measurement or Type 1 process creates distinctions so the natural characterization of the Type 2 processes would be ones that make no distinctions.

The extent to which two quantum states are indistinct or distinct is given by their inner product, i.e., their overlap. When their inner product is zero, then there is zero indistinctness or zero overlap between the states, i.e., they are fully distinct. Hence the natural characterization of the Type 2 processes as not changing the indistinctness or distinctness between quantum states would be a process that preserves inner products, i.e., a unitary transformation.

Hence the division of quantum processes into Type 1 and Type 2 is just the division between the processes that makes distinctions and those that don't. This meshes perfectly with Feynman's account (see below) of when to add probabilities (when alternative outcomes are distinguishable as in Type I processes) and when to add amplitudes (when alternative outcomes are indistinguishable as in Type II processes).

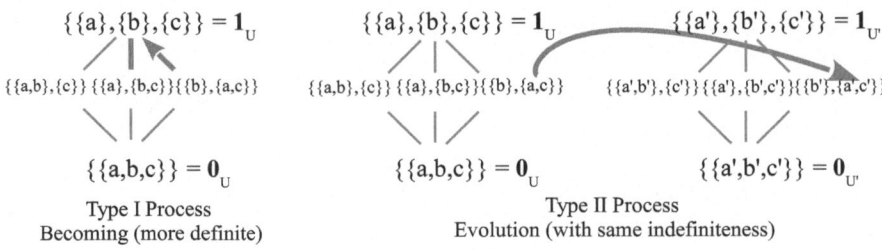

Fig. 4.1 *Anshaulich* (intuitive) picture of Type I and II processes represented by the 'skeletal' lattices of set partitions

What about the Schrödinger equation? The connection between unitary transformations and the solutions to the Schrödinger equation is given by Stone's Theorem (Stone 1932): there is a one-to-one correspondence between strongly continuous 1-parameter *unitary* groups $\{U_t\}_{t \in \mathbb{R}}$ and Hermitian operators H on the Hilbert space so that $U_t = e^{iHt}$ (solutions of the Schrödinger equation) (Hughes 1989, p. 114).

The difference between the two processes can also be illustrated at the set level in Fig. 4.1 where projective measurement is prefigured by vertical movement of the partition join and unitary evolution is prefigured by the horizontal movement of the transformation of one basis set $U = \{a, b, c\}$ into another basis set $U' = \{a', b', c'\}$ (as in the previous double-slit example).

Figure 4.1 allows us to answer the question: "Why there is only one fundamental process in classical mechanics (evolving definite states to definite states) and two fundamental processes in quantum mechanics." When is the Type I quantum notion of becoming (indefinite to more definite) not possible? The characteristic aspect of QM is superposition and at the level of fully definite classical states (discrete partition where there is no superposition), then there is no Type I process possible. There is only the 'horizontal' Type II process, as it were, of definite classical states evolving into definite classical states–as in classical mechanics.

Once the misleading wave-interpretation is replaced by the indefiniteness interpretation, then the "evolution of the wave function by the Schrödinger equation" is seen to be the unitary evolution of an indefinite state in one ON basis to an indefinite state in another ON basis. In schematic terms, it preserves the level of indefiniteness as in the horizontal evolution of the superposition $\{a, c\}$ into the superposition $\{a', c'\}$ in Fig. 4.1. This Fig. 4.1, summarizes at the skeletal level, the two fundamental non-classical processes, on the left side, the Type I process of jumping from an indefinite state to a more definite state (resulting from being in-formed by the distinctions of an observable), and, on the right side, the Type II process of unitary evolution.

One often sees a misleading complacency in citing unitary evolution as being like the classical deterministic evolution of classical physics and the resulting push to eliminate the Type I processes ("solve the measurement problem") to restore the more familiar determinism of classical physics. This is misleading since at the quantum level of reality, there are different levels of indefiniteness, and the unitary evolution and interference effects (e.g., in the two-slit experiment with no detection or distinctions at the slits) are because the unitary evolution is taking place at the non-classical level (as pictured in Fig. 4.1). Both types of processes are needed to describe quantum reality and are easily pictured at the skeletal level in the partition lattices of Fig. 4.1 which prefigure the Hilbert space versions in the math of QM. Figure 4.1 is thus a key figure to illustrate, in an *anshaulich* manner, some of the basic points in our analysis.

4.2.2 Measurement and the Collapse Postulate

We have seen that quantum measurements create distinctions. Richard Feynman was perhaps the quantum theorist who most emphasized measurement as making distinctions and, more generally, recognized that distinguishability and indistinguishability are key organizing concepts in QM.

> Feynman's approach is based on the contrast between processes that are *distinguishable* within a given physical context and those that are *indistinguishable* within that context. A process is distinguishable if some record of whether or not it has been realized results from the process in question; if no record results, the process is indistinguishable from alternative processes leading to the same end result. In my terminology, a registration of the realization of a process must exist for it to be a distinguishable alternative. In the two-slit experiment, for example, passage through one slit or the other is only a distinguishable alternative if a counter is placed behind one of the slits; without such a counter, these are indistinguishable alternatives. Classical probability rules apply to distinguishable processes. Nonclassical probability amplitude rules apply to indistinguishable processes. (Stachel 1986, p. 314)

The importance of the Feynman rules is that they reduce or bypass the whole hopeless discussion of "the measurement problem" in favor of rules where the operative concepts are distinguishability and indistinguishability (or distinctions and indistinctions). Since those concepts are the basic ones at the logical level (i.e., partition logic), that means the key interpretive concepts for our most fundamental physical theory (quantum mechanics), are indeed fundamental logical concepts. Given the unreasonable effectiveness of mathematical concepts in physics (Wigner 1967), this should not be surprising.

Logical information theory (Ellerman 2021) based on the notion of logical entropy, defines information as distinctions or 'distingishings'. Charles Bennett, one of the founders of quantum information theory, also saw information as "the notion of distinguishability abstracted away from what we are distinguishing, or from the carrier of information" (Bennett 2003, p. 155). Here Bennett meets Feynman to provide a possible explanation why so many quantum theorists think the "information" is somehow key to understanding and interpreting quantum mechanics.

When a superposition of eigenstates undergoes an interaction, is there a distinction made in principle between the superposed eigenstates in the interaction? If the eigenstates are distinguished by the interaction, then a state reduction takes place, the superposition is reduced (i.e., the so-called "wave function" collapses), and the probability of a later final state will add the probabilities (rather than amplitudes) of the eigenstates leading to the final state. If there is no differences or distinctions between the superposed eigenstates undergoing the interaction, then no measurement takes place and the amplitudes are added.

> If you could, *in principle*, distinguish the alternative *final* states (even though you do not bother to do so), the total, final probability is obtained by calculating the *probability* for each state (not the amplitude) and then adding them together. If you *cannot* distinguish the final states *even in principle*, then the probability amplitudes must be summed before taking the absolute square to find the actual probability. (Feynman et al. 1965, pp. 3–9)

For instance, we saw previously in the two-slit experiment how distinctions (detectors) at the slits meant adding probabilities and how no distinctions at the slits meant adding amplitudes. Anton Zeilinger makes the same point using the notion of information-as-distinguishability.

> In other words, the superposition of amplitudes ... is only valid if there is no way to know, even in principle, which path the particle took. It is important to realize that this does not imply that an observer actually takes note of what happens. It is sufficient to destroy the interference pattern, if the path information is accessible in principle from the experiment or even if it is dispersed in the environment and beyond any technical possibility to be recovered, but in principle still "out there." The absence of any such information is the essential criterion for quantum interference to appear. (Zeilinger 1999, p. 484)

What Zeilinger calls "information," Feynman calls "distinguishability," since information "is the notion of distinguishability abstracted away from what we are distinguishing, or from the carrier of information...." (Bennett 2003, p. 155). It might also be noted that Zeilinger and Brukner have argued that information should be measured by the formula for logical entropy which defines information as distinctions, differences, or distinguishability (Brukner and Zeilinger 2003).

Feynman, Stachel, Zeilinger, and others thus answer a question posed in the literature where the key concepts of distinguishability and indistinguishability (or distinctions and indistinctions) are ignored (as in most of the philosophy of quantum mechanics literature) and thus the difference between Type I and Type II processes may seem "unbelievable."

> It indeed seems necessary to admit that "measurements" are ubiquitous, and occur even in places and times where there are no human experimenters. But it also seems hopeless to think that we will be able to give an appropriately sharp answer to the question of what, exactly, differentiates the 'ordinary' processes (where the usual dynamical rules apply) from the 'measurement-like' processes (where the rules momentarily change). (Norsen 2017, p. 64)

> [I]t seems unbelievable that there is a fundamental distinction between " measurement" and "non-measurement" processes. Somehow, the true fundamental theory should treat all processes in a consistent, uniform fashion. (Norsen 2017, p. 245)

The "fundamental distinction" is between processes where distinctions are made or are not made between the eigenstates in the superposition. All the major interpretations of QM (Maudlin 2019; Norsen 2017) ignore Feynman's insistence in 1951 (Feynman 1951) on role of distinguishability or indistinguishability of alternative ways to go from A to B to determine whether a so-called "measurement" occurs (as Stachel put it, "In my terminology, a registration of the realization of a process must exist for it to be a distinguishable alternative" (Stachel 1986, p. 314) or does not occur.[2]

Feynman gives examples of measurement entirely at the quantum level, and thus he undercuts the long and tortured discussion about measurement as involving a macroscopic apparatus. When a particle scatters off the atoms in a crystal, the question of whether or not it should be treated as a superposition of scattering off the different atoms or as a mixture of scattering off of particular atoms with certain probabilities–hinges on distinguishability. If there was no distinction between scattering off different atoms, then no 'measurement' took place in the interaction and the superposition pure state evolves as a pure state. But if there was some distinction caused by scattering off an atom, then the result is the mixed state of scattering off the different atoms with different probabilities. If all the atoms had spin up and scattering off an atom flipped the spin, then a distinction was made so that constituted a measurement.

It should be noted that this and other examples of Feynman (Feynman et al. 1965; Feynman and Hibbs 1965, pp. 17–8) involve only quantum level interactions and thus have nothing to do with the unclear and shifty split ("Heisenberg cut") between microscopic and macroscopic, and thus are independent of the notion of "decoherence" based on interactions with macroscopic systems (e.g., Zurek 2003). A macroscopic apparatus for the use of humans must amplify the quantum level state reductions in order to be recorded, but such human level considerations should play no role in the theory of QM.

The difference between an interaction that constitutes a state reduction ("measurement") or not is whether or not any distinction exists between the different superposed eigenstates undergoing the interaction.[3] Feynman's implicit rule about state reduction might be paraphrased:

State Reduction Principle
If the interaction distinguishes between superposed eigenstates, then a state reduction (measurement) is made.

This is essentially a reformulation of Stachel's statement that a distinguishable alternative involves a "registration of the realization of a process," i.e., a state reduction. The underlying idea is simple; the eigenstates in a superposition are blurred together like elements in an equivalence class so when an interaction makes a distinction between them then the superposition is broken apart into a mixed state one part of which is probabilistically realized (according to the Born rule probabilities). The mathematics behind the State Reduction Principle can be formulated in both the set and Hilbert space cases.

---

[2] It is perhaps no surprise that Feynman is said to have quipped: "philosophy of science is about as useful to scientists as ornithology is to birds".

[3] In terms of our set level model, making a distinction between a and c in the superposition state $\{a, c\}$ transforms it into the mixed state $\{\{a\}, \{c\}\}$.

Proposition 4.2 (State Reduction Principle–set case) *A non-zero off-diagonal element $\rho(\pi)_{ik}$ (indicating a non-singleton superposition block in π) is zeroed (i.e., decohered) in the σ-measurement where $\sigma = g^{-1}$ for $g : U \to \mathbb{R}$ if and only if σ distinguishes between the elements u_i and u_k that were superposed in π, i.e., $g(u_i) \neq g(u_k)$.*

Proof The proof is of the contrapositive. An indistinction $(u_i, u_k) \in \text{indit}(\pi)$ remains an indistinction in $\rho(\pi \vee \sigma) = \sum_{j'=1}^{m'} P_{C_{j'}} \rho(\pi) P_{C_{j'}}$ where $\sigma = g^{-1} = \{C_1, ..., C_{m'}\}$ iff $(u_i, u_k) \in \text{indit}(\sigma)$, i.e., $g(u_i) = g(u_k)$. The result of the σ-measurement of $\rho(\pi)$ given by the Lüders mixture operation is $\pi \vee \sigma$ where $\text{indit}(\pi \vee \sigma) = \text{indit}(\pi) \cap \text{indit}(\sigma)$, so: if $(u_i, u_k) \in \text{indit}(\pi)$, then (u_i, u_k) is also an indistinction of $\pi \vee \sigma$ (i.e., the off-diagonal element $\rho(\pi \vee \sigma)_{ik} \neq 0$) iff $(u_i, u_k) \in \text{indit}(\sigma)$, i.e., $g(u_i) = g(u_k)$. □

The Hilbert space case follows, *mutatis mutandis*, from the set case. For the Hilbert space case, G is a Hermitian operator with a set of orthonormal eigenvectors $\{|u_i\rangle\}_{i=1}^{n}$ and a set $\{\lambda_j\}_{j=1}^{m}$ of real eigenvalues with corresponding eigenspaces $\{V_{\lambda_j}\}$. Let P_{V_λ} be the projection operator to the eigenspace V_λ for the eigenvalue λ. Then $P_{V_\lambda}|u_i\rangle = |u_i\rangle$ if $|u_i\rangle \in V_\lambda$, else 0 (zero vector), and since projection operators are Hermitian, $P_{V_\lambda}^{\dagger} = P_{V_\lambda}$ so we also have $\langle u_k| P_{V_\lambda} = \langle u_k|$ if $|u_k\rangle \in V_\lambda$, else 0.

Proposition 4.3 (State Reduction Principle–Hilbert space case) *A non-zero off-diagonal coherence element ρ_{ik} (indicating $|u_i\rangle$ and $|u_k\rangle$ are in superposition in ρ) is zeroed (decohered) in the G-measurement outcome $\hat{\rho} = \sum_\lambda P_{V_\lambda} \rho P_{V_\lambda}$ if and only if G distinguishes between $|u_i\rangle$ and $|u_k\rangle$, i.e., $|u_i\rangle$ and $|u_k\rangle$ have different eigenvalues.*

Proof The proof is of the contrapositive. A non-zero off-diagonal element ρ_{ik} remains the same in $\hat{\rho}_{ik}$ iff both $|u_i\rangle, |u_k\rangle \in V_\lambda$ for some V_λ, i.e., $|u_i\rangle$ and $|u_k\rangle$ have the same eigenvalue. An element ρ_{ik} can be written as $c_{ik}|u_i\rangle\langle u_k|$. In the Lüders mixture operation $\hat{\rho} = \sum_\lambda P_{V_\lambda} \rho P_{V_\lambda}$, the matrix $P_{V_\lambda} \rho$ is a row of column vectors $(P_{V_\lambda} \rho)_{.j}$ where the ith entry is $(P_{V_\lambda} \rho)_{ij} = c_{ij}|u_i\rangle\langle u_j|$ if $|u_i\rangle \in V_\lambda$, else 0. Then the entry $\hat{\rho}_{ik}$ is formed by the scalar product of the ith row $(P_{V_\lambda} \rho)_{i.}$ of $P_{V_\lambda} \rho$ times the kth column $(P_{V_\lambda})_{.k}$ of P_{V_λ}, so $\hat{\rho}_{ik} = \sum_{j=1}^{n} (P_{V_\lambda} \rho)_{ij} (P_{V_\lambda})_{jk} = c_{ik}|u_i\rangle\langle u_k| = \rho_{ik}$ iff both $|u_i\rangle, |u_k\rangle \in V_\lambda$, i.e., $|u_i\rangle$ and $|u_k\rangle$ have the same eigenvalue. □

If no distinctions are made, then no measurement or state reduction takes place.

One image for the measurement process is a 'shapeless' or indefinite blob of dough which then passes through a sieve or grating and acquires a definite polygonal shape as illustrated in Fig. 4.2 (a four-way "fork in the road"). The indefinite blob can be thought of as the superposition of the four definite shapes.

On the left side of Fig. 4.2, the interaction between the superposed blob and the sieve/grating forces a distinction, so a distinction is made as the blob must pass through one of the definite-shaped holes. In general, a state reduction ('measurement') from an indefinite superposition to a more definite state takes place when the particle in the superposition state undergoes an interaction that makes a distinction,

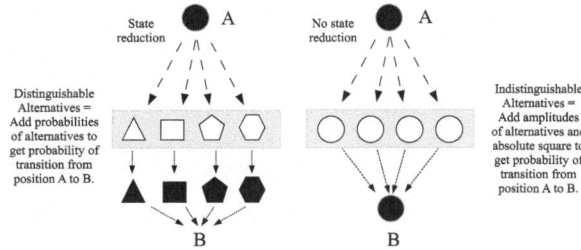

Fig. 4.2 Using grating imagery to illustrate Feynman's rule for distinguishable or indistinguishable alternatives

i.e., acts like a sieve as in Fig. 4.2 (left side). Then the probabilities of the four paths from A to B are added to get the total A-to-B probability.

If the interaction of the superposition with the grating does not distinguish between the superposed states (right side of Fig. 4.2), then no state reduction takes place, i.e., unitary evolution from position A to B. Then the amplitudes for getting from A to B by passing through the four holes in the 'null-grating' are added, and then the absolute square gives the total probability of going from A to B.

4.2.3 The Partition Logical Principle of Identity of Indistinguishables

The four eigen-shapes of Fig. 4.2 can be abbreviated as $\{a\}$, $\{b\}$, $\{c\}$, and $\{d\}$, and then we have the partition lattice with the classical world of fully distinct states at the top ('tip of the iceberg') and the quantum world of pure and mixed states including superposition states ('underwater part of the iceberg'). The fact that the discrete partition $\mathbf{1}_U$ represents the skeletal fully distinguished classical world is shown by the partition logic version of Leibniz's principle:

> Partition Logic Principle of Identity of Indistinguishables
> For all $u, u' \in U$, if $(u, u') \in \text{indit}(\mathbf{1}_U)$, then $u = u'$.

In the classical metaphysics of definite-all-the-way-down as expressed in Leibniz's identity of indiscernibles principle or Kant's principle of complete determination, all possible distinctions can be made. The only impossible distinctions are of an entity with itself–so if no distinction can be made, it is the self-same entity. In the discrete partition $\mathbf{1}_U$ on the universe set U, all the possible distinctions are made so the set of distinctions is $\text{dit}(\mathbf{1}_U) = U \times U - \Delta$ where diagonal Δ is the set of all self-pairs $\Delta = \{(u_1, u_1), ..., (u_n, u_n)\}$. Equivalently, the set of indistinctions for $\mathbf{1}_U$ is:

$$\text{indit}(\mathbf{1}_U) = \cup_{i=1}^n \{u_i\} \times \{u_i\} = \Delta = U \times U - \text{dit}(\mathbf{1}_U).$$

Hence the Principle of Identity of Indistinguishables holds for $\mathbf{1}_U$. In the quantum world of skeletal superposition states, i.e., non-singleton partition blocks, that

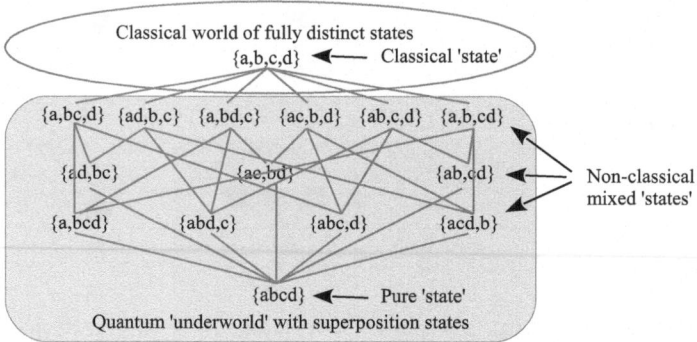

Fig. 4.3 The classical and 'quantum' parts in the skeletal partition lattice with one particle and four 'eigenstates'

principle is false, i.e., "If $(u_i, u_k) \in$ indit (π), then $u_i = u_k$" is true iff $\pi = \mathbf{1}_U$–so the Principle of Identity of Indistinguishables is equivalent to *classicality* in the lattice of partitions as illustrated in Fig. 4.3. The density matrix $\rho(\mathbf{1}_U)$ of the discrete partition is a diagonal matrix, and in the general quantum case, the 'classical' mixtures have diagonal density matrices, i.e., no coherences (no non-zero off-diagonal elements) (Auletta et al. 2009, p. 176).

4.2.4 Weyl's Partition Imagery

The sieve imagery of Fig. 4.2 has been used before. In his popular writing, Arthur Eddington used the sieve metaphor:

> In Einstein's theory of relativity the observer is a man who sets out in quest of truth armed with a measuring-rod. In quantum theory he sets out armed with a sieve. (Eddington 1947, p. 267)

Hermann Weyl cited that passage (Weyl 1949, p. 255) in his expositional concept of gratings. Weyl, in effect, used the Yoga of Linearization by taking *both* set partitions and vector space partitions (direct-sum decompositions) as the respective types of gratings (Weyl 1949, pp. 255–257). He started with a numerical attribute on a set, e.g., $f : U \to \mathbb{R}$, which defined the set partition or "grating" (Weyl 1949, p. 255) or "aggregate [which] is used in the sense of 'set of elements with equivalence relation.'" (Weyl 1949, p. 239) with blocks having the same attribute-value, e.g., $\{f^{-1}(r)\}_{r \in f(U)}$. Then he moved to the QM case where the "aggregate of n states has to be replaced by an n-dimensional Euclidean vector space" (Weyl 1949, p. 256). Then the notion of a vector space partition or "grating" in QM is a "splitting of the total vector space into mutually orthogonal subspaces" so that "each vector $\overrightarrow{x}$ splits into r component vectors lying in the several subspaces" (Weyl 1949, p. 256), i.e., a direct-sum decomposition of the space. After thus referring to a partition and a

DSD as a "grating" or "sieve," Weyl notes that "Measurement means application of a sieve or grating" (Weyl 1949, p. 259) (see Fig. 4.2), i.e., the making of distinctions by the join-like process described by the Lüders mixture operation.

4.2.5 Trajectories or Jumps?

Our overall goal has been to show that the mathematics of QM is the partition mathematics of distinctions and indistinctions or definiteness and indefiniteness expressed in terms of Hilbert spaces. The vision of realism based on objective indefiniteness is juxtaposed to our ordinary intuitive idea that reality is fully definite. One aspect of "the measurement problem" has been the lack of any mathematical description of the 'trajectory' of a quantum system going from an objectively indefinite superposition state to a more definite eigenstate during a measurement. But that question seems to arise out of imposing the fully-definite framework as if it was only a continuous transition from a definite state to another definite state with more discernible features. But if quantum reality consists of objectively indefinite states, why should we expect that sort of a continuous transition between states of indefiniteness–as opposed to the notion of a genuine quantum jump?

Leibniz's view of reality as being fully definite (expressed in his identity of indiscernibles and represented in skeletal form in the identity of indistinguishables for 1_U) is naturally associated with his "Principle of Continuity" (Auletta 2019, p. 7) also expressed as "*Natura non facit saltus*" (Nature does not make jumps) (Leibniz 1996, Bk. IV, Chap. XVI). In the opposite case of objective indefiniteness, we should not be surprised to find the opposite case of discontinuous change, i.e., quantum jumps. It was previously noted that the set version of a (maximal) measurement is given by a choice function that takes a non-empty subset to one of its elements. If the subset is not a singleton, then the choice function gives an indeterministic jump. If the subset is a singleton, i.e., maximally definite like in the blocks of 1_U, then there is no indeterministic jump which prefigures the measurement of an eigenstate to get that outcome deterministically with probability one.

4.3 Indistinguishability of Like Particles

4.3.1 The Implication of Reality Not Being Fully Definite

The existence of the maximal description CSCOs implies that there are not "distinctions all the way down."

> In quantum mechanics, however, identical particles are truly indistinguishable. This is because we cannot specify more than a complete set of commuting observables for each of the particles; in particular, we cannot label the particle by coloring it blue. (Sakurai and Napolitano 2011, p. 446)

For an illustration at the skeletal level of sets and partitions, think of two particles of the same type which have two possible states which we can take as heads h or tails t (instead of spin up or down). At the skeletal level of sets, the tensor product of the two one-particle Hilbert spaces is just the direct product of the set of 'eigenstates':

$$\{h, t\} \times \{h, t\} = \{(h, h), (h, t), (t, h), (t, t)\}.$$

In our previous Fig. 4.3, we illustrated the lattice of partitions on four eigenstates for one particle so we can adapt it for the lattice of possible (classical) states of two particles with two possible states with the replacements: $a = (h, h)$, $b = (h, t)$, $c = (t, h)$, and $d = (t, t)$ to obtain Fig. 4.4 (where the superposition $\{(h, t), (t, h)\}$ is abbreviated $(h, t)(t, h)$ and different blocks are separated with the semi-colon).

But due to the "identity" of the like particles, there is no fact-of-the-matter difference between a state and one resulting from a permutation of the particles. Hence all the states that are not permutation invariant such as the superposition $\{(h, h), (h, t)\}$ (abbreviated $(h, h)(h, t)$ in the mixed states $\{(t, h); (t, t); (h, h)(h, t)\}$ or $\{(h, h)$ $(h, t); (t, h)(t, t)\}$) are not possible physical states so they are marked with "X" in Fig. 4.5.

The lattice of permitted states is given in Fig. 4.6 (where repetitions like (h, h) are allowed so these are boson states).

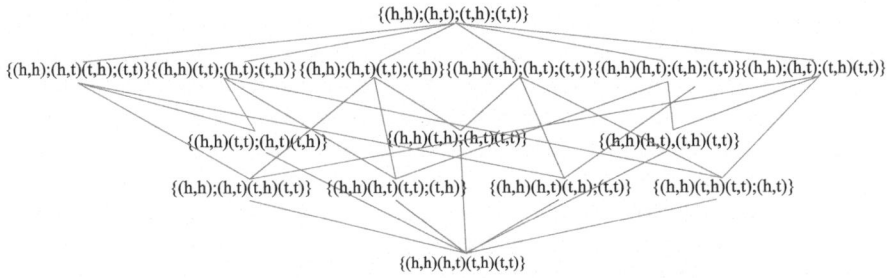

Fig. 4.4 Lattice of partitions of (classically) possible states (2 particles; 2 states)

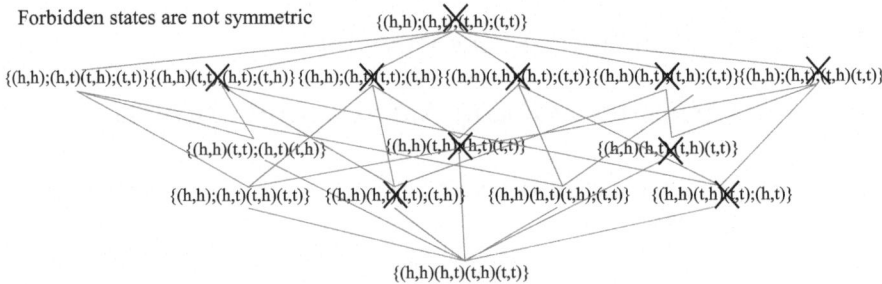

Fig. 4.5 States not permutation-invariant crossed out

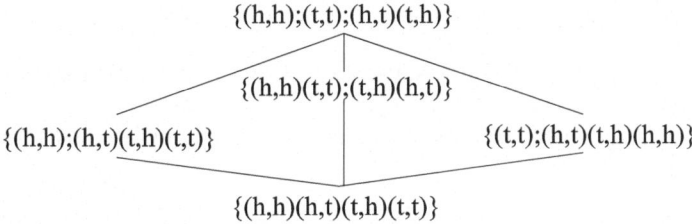

Fig. 4.6 Symmetric states for bosons

Note that the top of the lattice is no longer the discrete partition on the four classical states, but contains the superposition $\{(h, t)\, (t, h)\}$ since with like particles, there are no physical states such as (h, t) or (t, h). Nature cannot become more definite. Those three states in the mixture $\{(h, h)\,;\, (t, t)\}\,;\, (h, t)\, (t, h)\}$ are the most definite states; the remaining indefiniteness in the superposition $\{(h, t)\, (t, h)\}$ is objective. Fermions might be characterized as not allowing repetitions like (h, h) or (t, t), so the diagram like Fig. 4.6 but for fermions would have only one physical state, the superposition $\{(h, t)\, (t, h)\}$.

4.3.2 The Combinatorics of Fermions and Bosons

What is the source of the difference between fermions and bosons? The analysis (in terms of symmetric versus anti-symmetric states) cannot be prefigured by the skeletal set partitions since the elements from U in a partition have no scalars times them; the scalars come with linearization.

If quantum reality was definite all the way down, there would no such distinction; the Maxwell-Boltzmann statistics would prevail. If quantum reality is not definite all the way down, then at a certain level further distinctions stop. The *physics* of bosons versus fermions is based on integral spin (symmetric wave functions) or half-integral spin (anti-symmetric wave functions), but we might expect such a division just based on reasoning about definiteness and indefiniteness.

A metaphor would be the level of definiteness in a mailing address that was definite only down to the street number. A 'fermionic' neighborhood could be zoned so there is at most one resident per street number in suburban sprawl, e.g., only a single-family dwelling at the street number or not (vacant lot). But in a 'bosonic' residential district of a city, there might be multifamily dwellings (e.g., apartment houses) at each street number. If addresses could make further distinctions (e.g., according to apartment number), like painting numerically distinct particles of the same type by different colors, then there would again be at most one resident per address. But at the quantum level, those further distinctions might not be possible since it is not definite all the way down. Hence, in that case, there could be multiple residents (particles) at each street address (CSCO-defined state). Those two possibilities are present when addresses are not arbitrarily refineable.

In QM, those two possibilities are:

- fermions: a complete state description (e.g., CSCO-defined) is sufficient to limit at most a single particle to that state, or
- bosons: the complete description is still insufficient to limit the number of particles in that state.

The mathematics of the different quantum statistics for fermions and bosons can be developed using the standard combinatorics of balls in boxes–which brings out the underlying role of distinctions and indistinctions, unlike the usual treatment involving symmetric and anti-symmetric wave functions.

- Fermi-Dirac statistics is based on the number of ways indistinguishable balls (particles) are allocated to distinguishable boxes (states) using *distinction-preserving* (i.e., one-to-one) functions so two numerically distinct balls have to go to distinct boxes (i.e., repetitions are not allowed in a box), while:
- Bose-Einstein statistics is based on the number of ways indistinguishable balls (particles) are allocated to distinguishable boxes (states) using *arbitrary* functions and thus allowing many balls (particles) in the same box (state).

Intuitively, as illustrated in Fig. 4.7, fermions 'fill' a state whereas a boson in a state creates new 'slots' before and after it, so with equiprobable *slots*, bosons will tend to aggregate in a state (Feller 1968, p. 40).

With k particles of the same type, the k-fold tensor product of the Hilbert spaces of n possible particle states, there is a subspace of physically realizable states. Putting each fermion-ball into a box/state takes away that box as a choice for the next fermion to be distributed, while putting a boson-ball into a box/state adds a choice (before or after that boson) for the next boson to be distributed, and then, in both cases, the division by $k!$ (to cancel out the ordering of choices) computes the respective number of basis vectors (dimension of the subspace). The difference between the two dimensions is given by the difference between the *falling factorial* $n (n - 1) ...(n - k + 1)$ (k terms) and the *rising factorial* $n (n + 1) ...(n + k - 1)$ (k terms).

Thus the dimension of the bosonic subspace is:

$$\left\langle \frac{n}{k} \right\rangle = \frac{n (n + 1) ...(n + k - 1)}{k!}.$$

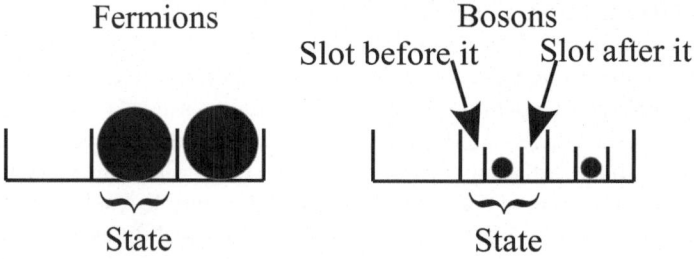

Fig. 4.7 Some imagery for fermions and bosons

The Bose-Einstein statistics counts each state as having equal probability $1/\left\langle{n \atop k}\right\rangle$. In the previous example of the set lattice with two particles and two states, $\left\langle{2 \atop 2}\right\rangle = \frac{2(3)}{2!} = 3$ which was the number of 'eigenstates,' i.e., $\{(h, h)\}$, $\{(t, t)\}$, and $\{(h, t) (t, h)\}$.

The dimension of the fermionic subspace is the usual binomial coefficient:

$$\binom{n}{k} = \frac{n(n-1)\ldots(n-k+1)}{k!} = \frac{n(n-1)\ldots(n-k+1)(n-k)!}{k!(n-k)!} = \frac{n!}{k!(n-k)!}.$$

The Fermi-Dirac statistics counts each state as having equal probability: $1/\binom{n}{k}$. In the previous example, when the states with repetitions, i.e., $\{(h, h)\}$ and $\{(t, t)\}$, are removed, then only the $\binom{2}{2} = 1$ state remained, namely $\{(h, t) (t, h)\}$.

In the classical (i.e., fully distinguishability) case of Maxwell-Boltzmann (MB) statistics, the k balls (particles) are distinguishable, the n boxes (states) are distinguishable, and the distributions are by arbitrary functions. There are $k!$ linear orders (or permutations) of the k distinguishable particles but they are grouped into n boxes with the occupation numbers of $\theta_1, \ldots, \theta_n$ for the n boxes. How many distributions are there with those occupation numbers? It is perhaps surprising that the answer is still $k!$ independent of the occupation numbers. The proof is illustrated in Fig. 4.8 since the $n-1$ 'walls' or dividers to make the boxes can be put in arbitrarily.

But the ordering of the balls within each box does not matter so we need to divide through by the $\theta_i!$ for $i = 1, \ldots, n$ to get the total number of possible states with those occupation numbers for the distinguishable boxes (with the ordering $1, \ldots, n$). The result is the well-known *multinomial coefficient*:

$$\binom{k}{\theta_1, \ldots, \theta_n} = \frac{k!}{\theta_1!\ldots\theta_n!}.$$

There are n^k arbitrary functions distributing the balls in the boxes and each distribution is classically considered equiprobable so the probability of the given set of occupation numbers in the MB statistics is:

$$\binom{k}{\theta_1, \ldots, \theta_n}/n^k = \frac{k!}{\theta_1!\ldots\theta_n!}/n^k.$$

In the previous example, when permutation symmetry was not required, there were four 'eigenstates' (h, h), (h, t), (t, h), (t, t) which form a basis for the

Fig. 4.8 Number of ways to fill k boxes with given occupation numbers is $k!$

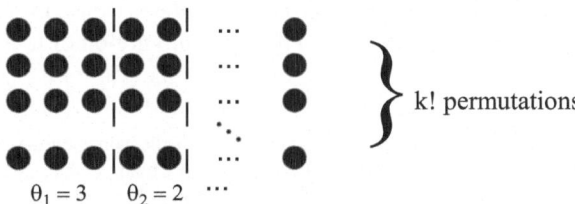

four-dimensional space $\mathbb{C}^2 \otimes \mathbb{C}^2$ with the three-dimensional subspace of possible boson states and the one-dimensional subspace of possible fermion states.

Continuing the example to illustrate the difference between MB, BE, and FD statistics, let us compute the probability of flipping two coins of the same type and getting the state of different outcomes. Hence there are $k = 2$ particles of the same type and $n = 2$ states $\{h, t\}$ like two coins with heads and tails as the states. What is the probability that one "coin" will be "heads" and the other "tails"?

- Classical coins: $\Pr_{MB}(\{(h, t), (t, h)\}) = \frac{\binom{k}{\theta_1, \dots, \theta_n}}{n^k} = \frac{\frac{k!}{\theta_1! \dots \theta_n!}}{n^k} = \frac{2}{4} = \frac{1}{2}$.
- Boson coins: $\Pr_{BE}(\{(h, t), (t, h)\}) = \frac{k!}{n(n+1) \dots (n+k-1)} = \frac{2}{2(3)} = \frac{1}{3}$.
- Fermion coins: $\Pr_{FD}(\{(h, t), (t, h)\}) = \frac{k!}{n(n-1) \dots (n-k+1)} = \frac{2}{2(1)} = 1$.

The two fermion coins *have* to be in different states (i.e., with probability 1) which illustrates the Pauli exclusion principle. In the bosonic case, the two classical outcomes (h, t) and (t, h) differ only by a permutation of particles of the same type so that counts only as one state out of three equiprobable states. The skeletal partition-version of the 3-dimensional bosonic subspace was illustrated in Fig. 4.6. The probability of getting the same outcomes (h, h) or (t, t) is $\frac{2}{3}$ in the bosonic case in comparison with the classical MB probability of $\frac{1}{2}$ which illustrates the "social" tendency of bosons to "want" to be in the same state (e.g., live in an apartment house in our metaphorical terms).

4.4 Measurement with Quantum Logical Entropy

4.4.1 From Logical Entropy to Quantum Logical Entropy

This section deals with the notion of entropy, quantum logical entropy, that (unlike von Neumann entropy) is computed from the quantum coherences (non-zero entries in the density matrix) that are decohered (i.e., zeroed) in a projective measurement–just as in the set case, the logical entropy was computed from the indistinctions that were distinguished in the application of the Lüders mixture operation (set version). This is a case where the Yoga is used to develop the appropriate quantum mathematics that meshes precisely (down to the level of indistinctions $\approx$ coherences) with the machinery of projective measurement, i.e., the set and quantum versions of the Measuring Measurement Theorems.[4]

Using the Yoga of Linearization, the set-based notion of logical entropy for set partitions easily extends to the quantum notion of logical entropy.

---

[4] There is no such precise relationship with the von Neumann entropy even though it increases under projective measurement (Nielsen and Chuang 2000, p. 515)–which is not surprising since the set notion of Shannon entropy is a nonlinear but monotonic transform (the dit-bit transform) of logical entropy (Ellerman 2021).

- The sets notion of logical entropy was defined on set partitions determined by numerical attributes $f : U \to \mathbb{R}$; and
- the corresponding quantum notion of logical entropy is given by DSDs determined by Hermitian operators $F : V \to V$ on a Hilbert space V which have an ON (orthonormal) basis U of eigenvectors with an eigenvalue function $f : U \to \mathbb{R}$ where $F u_i = f(u_i) u_i$.

In a similar manner, the notion of information-as-distinctions extends to the quantum domain.

- An ordered pair of set elements (u_i, u_k) in the direct product $U \times U$ is a distinction or dit if they have different numerical attributes $f(u_i) \neq f(u_k)$, and
- an ordered pair of eigenvectors written $u_i \otimes u_k$ in the tensor product $V \otimes V$ is a *qudit* if they have different eigenvalues $f(u_i) \neq f(u_k)$.[5]

4.4.2 Quantum Logical Entropy Without Probabilities

Both the set version and quantum versions of logical entropy satisfy Andrei Kolmogorov's dictum:

> Information theory must precede probability theory, and not be based on it. By the very essence of this discipline, the foundations of information theory have a finite combinatorial character. (Kolmogorov 1983, p. 39)

In the set version, logical entropy [unlike Shannon entropy] is represented by the finite combinatorial ditset dit $(\pi) \subseteq U \times U$ without any assumed probability distribution, and, the quantum logical entropy (unlike von Neumann entropy $S(\rho) = -\operatorname{tr}[\rho \ln(\rho)]$) is represented by the finite dimensional subspace $\left[\text{qudit}(F)\right] \subseteq V \otimes V$ (the subspace generated by the set qudit(F) of qudits $u_i \otimes u_k$ of F) without any assumed density matrix supplying the probabilities. Table 4.2 gives the without-probabilities translation dictionary between the set level logical entropy and the quantum logical entropy.

There is an interesting theorem about dits that carries over to qudits. The ditset of the indiscrete partition $\mathbf{0}_U$ is empty, but there is a Common Dits Theorem (Ellerman 2009) that any two non-empty ditsets dit (π) and dit (σ) have a non-empty intersection for any partitions π and σ on the same set U. This fact is usually stated in its equivalence relation form: if the union of two equivalence relations indit (π) and indit (σ) is the universal equivalence relation $U \times U$, then one of the two equivalence relations is already the universal one. Or, if $\emptyset \neq$ dit $(\pi) \subseteq$ indit (σ), then $\sigma = \mathbf{0}_U$. In graph theory, it is the fact that if a simple graph G is disconnected, then its complementary graph G^c (same nodes but only the links not in G) is connected (Wilson 1972, p. 30).

---

[5] This use of the word "qudit" to denote a QUantum DIT is quite different from its use just to describe a d-dimensional quantum system, e.g., (Bertlmann and Krammer 2008).

Table 4.2 Ditsets and qudit subspaces without probabilities

Classical logical information	Quantum logical information
$f, g : U \to \mathbb{R}$	Commuting Hermitian ops. F, G
$U = \{u_1, \ldots, u_n\}$	ON basis simultaneous eigenvectors F, G
Values $\{\phi_i\}_{i=1}^{m}$ of f	Eigenvalues $\{\phi_i\}_{i=1}^{m}$ of F
Values $\{\gamma_j\}_{j=1}^{m'}$ of g	Eigenvalues $\{\gamma_j\}_{j=1}^{m'}$ of G
Partition $\{f^{-1}(\phi_i)\}_{i=1}^{m}$	Eigenspace DSD of F
Partition $\{g^{-1}(\gamma_j)\}_{j=1}^{m'}$	Eigenspace DSD of G
dits $\pi : (u_k, u_{k'}) \in U^2$, $f(u_k) \neq f(u_{k'})$	Qudits F: $u_k \otimes u_{k'} \in V \otimes V$, $f(u_k) \neq f(u_{k'})$
dits $\sigma : (u_k, u_{k'}) \in U^2$, $g(u_k) \neq g(u_{k'})$	Qudits G: $u_k \otimes u_{k'} \in V \otimes V$, $g(u_k) \neq g(u_{k'})$
dit $(\pi) \subseteq U \times U$	$\left[\text{qudit}(F)\right]$ = subspace gen. by qudits of F
dit $(\sigma) \subseteq U \times U$	$\left[\text{qudit}(G)\right]$ = subspace gen. by qudits of G
dit $(\pi) \cup$ dit $(\sigma) \subseteq U \times U$	$\left[\text{qudit}(F) \cup \text{qudit}(G)\right] \subseteq V \otimes V$
dit $(\pi) -$ dit $(\sigma) \subseteq U \times U$	$\left[\text{qudit}(F) - \text{qudit}(G)\right] \subseteq V \otimes V$
dit $(\pi) \cap$ dit $(\sigma) \subseteq U \times U$	$\left[\text{qudit}(F) \cap \text{qudit}(G)\right] \subseteq V \otimes V$

Carried over to qudits, this means that given two commuting non-constant observables F and G with simultaneous eigenvectors $U = \{u_1, \ldots, u_n\}$, there is always a pair of simultaneous eigenvectors $u_i \neq u_k$ that have different eigenvalues for *both* F and G. Simply apply the Common Dits Theorem to the two partitions obtained as the inverse-image of the two eigenvalue functions $f, g : U \to \mathbb{R}$ for F and G so $f(u_i) \neq f(u_k)$ and $g(u_i) \neq g(u_k)$.

4.4.3   Quantum Logical Entropy with Probabilities

In the set case of logical entropy, we assumed a point probability distribution p on U and then applied the product probability distribution $p \times p$ to the ditset dit $\left(f^{-1}\right)$ of the inverse-image partition $\left\{f^{-1}(\phi_i)\right\}_{i=1}^{m}$ to get the logical entropy $h\left(f^{-1}\right) = p \times p\left(\text{dit}\left(f^{-1}\right)\right)$ which was interpreted as the probability of getting different f-values in two independent samples of the random variable f. In short, the logical entropy of a partition is the product probability measure's value on the ditset of the partition.

In the quantum or vector space case, the probabilities only enter by considering a certain state $\psi = \sum_i \alpha_i u_i$ with a density matrix

$$\rho(\psi)_{ij} = (|\psi\rangle \langle\psi|)_{ij} = \left(\left[\cdots, \alpha_i, \cdots\right]^t \left[\cdots, \alpha_j^*, \cdots\right]\right) = \alpha_i \alpha_j^*$$

(represented in the basis U of F eigenvectors) and then the density matrix $\rho(\psi) \otimes \rho(\psi)$ on $V \otimes V$. The probability of getting distinct eigenvalues in two independent measurements of the same prepared state ψ by the observable F is the trace of the projection $P_{[\text{qudit}(F)]}$ to the qudit space $[\text{qudit}(F)]$ times the density matrix $\rho(\psi) \otimes \rho(\psi)$, and that is the quantum logical entropy associated with the F-measurement of ψ:

$$h(F : \psi) := \text{tr}\left[P_{[qudit(F)]}\rho(\psi) \otimes \rho(\psi)\right].$$

We previously saw that logical probability and logical entropy had respective one-draw and two-draw probability interpretations. This extends to the quantum case. The quantum probability $\Pr(\phi) = \text{tr}\left[P_\phi \rho(\psi)\right]$ (where P_ϕ is the projection operator to the eigenspace V_ϕ of the eigenvalue ϕ and $\rho(\psi)$ is the density matrix of state ψ represented in measurement basis) is the one F-measurement probability of getting the eigenvalue ϕ. Similarly, the quantum logical entropy $h(F : \psi) = \text{tr}\left[P_{[qudit(F)]}\rho(\psi) \otimes \rho(\psi)\right]$ is the probability in two independent F-measurements of the same state ψ of getting different eigenvalues. The tensor product $\rho(\pi) \otimes \rho(\pi)$ is an $n^2 \times n^2$ matrix with the diagonal entries $(\rho(\pi) \otimes \rho(\pi))_{(j,k),(j,k)} = \rho(\pi)_{jj}\rho(\pi)_{kk} = p_j p_k$.

We also previously saw that logical entropy for set partitions could also be defined in the density matrix formulation as $h(\rho(\pi)) = 1 - \text{tr}\left[\rho(\pi)^2\right] = h(\pi)$. This suggests the simple formula $h(\rho) = 1 - \text{tr}\left[\rho^2\right]$ for the notion of quantum logical entropy of an arbitrary density matrix ρ independent of any measurement (Ellerman 2022).[6]

An arbitrary density matrix ρ has a set of non-negative real eigenvalues $\rho_1, ..., \rho_n$ which sum to one–so it qualifies as a probability distribution (with some zero probabilities). But there seems to be little attempt in the literature to assign an interpretation to these eigenvalues.

It is a tempting (and surprisingly common) fallacy to suppose that the eigenvalues and eigenvectors of a density matrix have some special significance with regard to the ensemble of quantum states represented by that density matrix. (Nielsen and Chuang 2000, p. 103)

However, it was previously shown that for density matrices of the special form $\rho(\pi)$ for a set partition $\pi = \{B_1, ..., B_m\}$ on U, the eigenvalues $\rho_1, ..., \rho_n$ are the block probabilities $\Pr(B_j)$ with zeros to fill out the $n - m$ eigenvalues. Since $\sum_{i=1}^n \rho_i^2 = \text{tr}\left[\rho^2\right]$ (Fano 1957, p. 77) this gives a standard interpretation to the eigenvalues $\{\rho_i\}$ of an arbitrary density matrix ρ in terms of the special density matrices $\rho(\pi)$ derived from a partition π on U. That is, the m non-zero eigenvalues ρ_j of ρ can be constructed as the block probabilities $\Pr(B_j)$ of a set partition π which has the same quantum logical entropy:

$$h(\rho) = 1 - \text{tr}\left[\rho^2\right] = 1 - \sum_{\rho_j \neq 0} \rho_j^2 = 1 - \sum_{j=1}^m \Pr(B_j)^2 = h(\pi)$$

where $\rho_j = \Pr(B_j)$ for $j = 1, ..., m$. Moreover, the spectral decomposition of an arbitrary density matrix ρ with eigenvalues $\{\rho_i\}$ and orthonormal eigenvectors $\{|\psi_i\rangle\}$

---

[6] This density matrix formula for logical entropy has been called "mixedness" (Jaeger 2007, p. 5).

shows that the density matrix can always be represented as constructed from *orthogonal* pure states (instead of *disjoint* blocks of a partition): $\rho = \sum_i \rho_i |\psi_i\rangle \langle\psi_i|$.

The skeletal set version of a pure state is the indiscrete partition $\mathbf{0}_U$ with only one block U of probability 1, and any pure quantum state's density matrix $\rho(\psi) = |\psi\rangle \langle\psi|$ has only one non-zero eigenvalue of 1. The indiscrete partition $\mathbf{0}_U$ does not make any distinctions so its logical entropy is 0 and its density matrix $\rho(\mathbf{0}_U)$ is idempotent. In the quantum case, a pure state has an idempotent density matrix $\rho^2 = \rho$ so $h(\rho) = 1 - \text{tr}[\rho^2] = 0$ for pure states–which also recalls the comparison between the indiscrete partition on a set U and the pure superposition of all the eigenstates U in the quantum case.

At the other extreme is the classical discrete partition $\mathbf{1}_U$ with equiprobable points so the block probabilities are all $\frac{1}{n}$ with the logical entropy $1 - \frac{1}{n}$ (the probability that the second draw from U differs from the first draw). The discrete partition with equiprobable points is the skeletal version of the maximally decomposed mixed quantum state that has a diagonal density matrix $\rho = \frac{1}{n}I$ with the diagonal entries and

Table 4.3 Probabilities applied to ditsets and qudit spaces (where $\rho(\psi)\rho(\psi) = \rho(\psi) \otimes \rho(\psi)$)

'Classical' logical entropy	Quantum logical entropy
$\text{Pr}(\phi_i) = \text{tr}\left[P_{f^{-1}(\phi_i)}\rho\left(f^{-1}\right)\right]$	$\text{Pr}(\phi_i) = \text{tr}\left[P_{\phi_i}\rho(\psi)\right]$
$\text{Pr}(\phi_i) = $ one-draw prob. of ϕ_i	$\text{Pr}(\phi_i) = $ one-measurement prob. of ϕ_i
Pure $\mathbf{0}_U$ with density matrix, $\rho(\mathbf{0}_U)$	Pure state ψ with density matrix $\rho(\psi) = |\psi\rangle \langle\psi|$
$U = \{u_1, ..., u_n\}$	ON basis simultaneous eigenvectors F, G
$p \times p$ on $U \times U$	$\rho(\psi) \otimes \rho(\psi)$ on $V \otimes V$
$h(\mathbf{0}_U) = 1 - \text{tr}\left[\rho(\mathbf{0}_U)^2\right] = 0$	$h(\rho(\psi)) = 1 - \text{tr}\left[\rho(\psi)^2\right] = 0$
$h(\mathbf{1}_U) = 1 - \frac{1}{n}\ (p_i = \frac{1}{n})$	$h\left(\frac{1}{n}I\right) = 1 - \text{tr}\left[\left(\frac{1}{n}I\right)^2\right] = 1 - \frac{1}{n}$
$h(\pi) = p \times p\ (\text{dit}(\pi))$	$h(F : \psi) = \text{tr}\left[P_{[\text{qudit}(F)]}\rho(\psi) \otimes \rho(\psi)\right]$
$h(\pi, \sigma) = p \times p\ (\text{dit}(\pi) \cup \text{dit}(\sigma))$	$h(F, G : \psi) = \text{tr}\left[P_{[\text{qudit}(F)\cup\text{qudit}(G)]}\rho(\psi)\rho(\psi)\right]$
$h(\pi|\sigma) = p \times p\ (\text{dit}(\pi) - \text{dit}(\sigma))$	$h(F|G : \psi) = \text{tr}\left[P_{[\text{qudit}(F)-\text{qudit}(G)]}\rho(\psi)\rho(\psi)\right]$
$m(\pi, \sigma) = p \times p\ (\text{dit}(\pi) \cap \text{dit}(\sigma))$	$m(F, G : \psi) = \text{tr}\left[P_{[\text{qudit}(F)\cap\text{qudit}(G)]}\rho(\psi)\rho(\psi)\right]$
$h(\pi) = h(\pi|\sigma) + m(\pi, \sigma)$	$h(F : \psi) = h(F|G : \psi) + m(F, G : \psi)$
$h(\pi) = $ 2-draw prob. diff. f-values	$h(F : \psi) = $ 2-meas. prob. diff. F-eigenvalues
$\rho(\pi) = \sum_i P_{B_i}\rho(\mathbf{0}_U)P_{B_i}$	$\hat{\rho}(\psi) = \sum_i P_{\phi_i}\rho(\psi)P_{\phi_i}$
$h(\pi) = 1 - \text{tr}\left[\rho(\pi)^2\right]$	$h(F : \psi) = 1 - \text{tr}\left[\hat{\rho}(\psi)^2\right]$
$h(\pi) = $ sum sq. zeroed $\rho(\mathbf{0}_U) \rightsquigarrow \rho(\pi)$	$h(F : \psi) = $ sum ab. sq. zeroed $\rho(\psi) \rightsquigarrow \hat{\rho}(\psi)$

eigenvalues $\frac{1}{n}$ so the quantum logical entropy is also $h\left(\frac{1}{n}I\right) = 1 - \text{tr}\left[\left(\frac{1}{n}I\right)^2\right] = 1 - \frac{1}{n}$ (Fano 1957, p. 84). In general, density matrices of the form $\rho\left(\pi\right)$ thus play a role in interpreting the eigenvalues of arbitrary $n \times n$ density matrices ρ.

Table 4.3 continues the definitions of quantum logical entropy by bring in probabilities where F and G are commuting observables so they have a basis of simultaneous eigenvectors (which is the analogue of two numerical attributes $f, g : U \to \Bbbk$ defined on the same set U). It is the with-probabilities translation dictionary between set level logical entropy and quantum logical entropy.

The last three lines of Table 4.3 anticipate the quantum version of the previous results about real-valued density matrices. One of the main results about density matrices (over the complex numbers where $\left\lVert\rho_{ij}\right\rVert^2$ is the absolute square of ρ_{ij}) is.

Proposition 4.4 $\text{tr}\left[\rho^2\right] = \sum_{i,j}\left\lVert\rho_{ij}\right\rVert^2$ *(Fano 1957, p. 77).*

Proof A diagonal entry in ρ^2 is $\left(\rho^2\right)_{ii} = \sum_{j=1}^{n}\rho_{ij}\rho_{ji}^* = \sum_{j=1}^{n}\left\lVert\rho_{ij}\right\rVert^2$ so $\text{tr}\left[\rho^2\right] = \sum_{i=1}^{n}\left(\rho^2\right)_{ii} = \sum_{i,j}\left\lVert\rho_{ij}\right\rVert^2$. $\qquad\square$

In general the quantum logical entropy of a density matrix ρ is: $h\left(\rho\right) = 1 - \text{tr}\left[\rho^2\right] = 1 - \sum_i \rho_i^2 = 1 - \sum_{ij}\left\lVert\rho_{ij}\right\rVert^2$ which corresponds to the logical entropy of a probability distribution $p = (p_1, ..., p_n)$ as $h\left(p\right) = 1 - \sum_{i=1}^{n}p_i^2$. That 'classical' logical entropy $h\left(p\right)$ corresponds to the logical entropy of a diagonal density matrix representing a completely decomposed mixture. The common theme in both $h\left(\rho\right) = 1 - \text{tr}\left[\rho^2\right] = 1 - \sum_{ij}\left\lVert\rho_{ij}\right\rVert^2$ and $h\left(p\right) = 1 - \sum_{i=1}^{n}p_i^2$ is "one minus the sum of probabilities of indistinctions $\sum_{i=1}^{n}p_i^2$ (or coherences $\sum_{i,j}\left\lVert\rho_{ij}\right\rVert^2$)"–which is the probability of distinctions.

The change in density matrices due to a projective measurement is given by the Lüders mixture operation. If V_i is the eigenspace for the eigenvalue ϕ_i of F and P_{ϕ_i} is the projection matrix $P_{\phi_i} : V \to V$ to that subspace, then the post-measurement density matrix is:

$$\hat{\rho}\left(\psi\right) = \sum_i P_{\phi_i}\rho\left(\psi\right)P_{\phi_i}.$$

As we saw previously in the set case, the Lüders mixture operation is like the join operation on partitions since it preserves only the non-zero off-diagonal elements of $\rho\left(\psi\right)$ that are also 'indistinctions' of F (i.e., that are non-qudits in the sense of having the same F eigenvalues)–which at the set level is just the intersection of equivalence relations.

The other previous result about the sum of the squares of the non-zero off-diagonal elements of ρ that are zeroed in the transition $\rho \rightsquigarrow \hat{\rho}$ carries over to the quantum case of density matrices over the complex numbers.

Theorem 4.1 (Measuring Measurement) *The increase in quantum logical entropy, $h\left(\hat{\rho}\left(\psi\right)\right)$ due to the F-measurement of the pure state ψ is the sum of the absolute squares of the non-zero off-diagonal terms (coherences $\sim$ indistinctions) in $\rho\left(\psi\right)$*

(represented in a basis of F-eigenvectors) that are zeroed ('decohered', i.e., turned into quantum distinctions) in the post-measurement Lüders mixture density matrix $\hat{\rho}(\psi) = \sum_i P_{\phi_i} \rho(\psi) P_{\phi_i}$.

Proof

$$h\left(\hat{\rho}(\psi)\right) - h\left(\rho(\psi)\right) = \left(1 - \operatorname{tr}\left[\hat{\rho}(\psi)^2\right]\right) - \left(1 - \operatorname{tr}\left[\rho(\psi)^2\right]\right) = \sum_{j,k}\left(\|\rho_{jk}(\psi)\|^2 - \|\hat{\rho}_{jk}(\psi)\|^2\right)$$

since $\operatorname{tr}\left[\rho^2\right] = \sum_{i,j}\|\rho_{ij}\|^2$ is the sum of the absolute squares of all the elements of ρ. If u_j and u_k are a qudit of F, then and only then are the corresponding non-zero off-diagonal terms in $\rho(\psi)$ zeroed by the Lüders mixture operation $\sum_{i=1}^{I} P_{\phi_i} \rho(\psi) P_{\phi_i}$ to obtain $\hat{\rho}(\psi)$ from $\rho(\psi)$. $\square$

A tedious calculation shows that $h(F : \psi) = h\left(\hat{\rho}(\psi)\right)$ which also equals the sum of the absolute squares of zeroed terms in the transition $\rho(\psi) \rightsquigarrow \hat{\rho}(\psi)$ (since pure states $\rho(\psi)$ have zero quantum logical entropy). These equalities are best illustrated by working through a simple example of measuring z-axis spin.

Example: Let $|\psi\rangle = \alpha_\uparrow |\uparrow\rangle + \alpha_\downarrow |\downarrow\rangle = \begin{bmatrix} \alpha_\uparrow \\ \alpha_\downarrow \end{bmatrix}$ be a pure normalized superposition state of z-spin up and z-spin down so the density matrix is $\rho(\psi) = \begin{bmatrix} p_\uparrow & \alpha_\uparrow \alpha_\downarrow^* \\ \alpha_\downarrow \alpha_\uparrow^* & p_\downarrow \end{bmatrix}$ (where α^* is the complex conjugate of α). For the observable F, let the eigenvalue function be $f : \{|\uparrow\rangle, |\downarrow\rangle\} \to \{+1, -1\}$ where $f(|\uparrow\rangle) = 1$ and $f(|\downarrow\rangle) = -1$. Then $F : \mathbb{C}^2 \to \mathbb{C}^2$ is represented by the matrix $F = \begin{bmatrix} 1 & 0 \\ 0 & -1 \end{bmatrix}$. The tensor product $\rho(\psi) \otimes \rho(\psi)$ is the $2^2 \times 2^2$ matrix:

$$\begin{bmatrix} p_\uparrow \rho(\psi) & \alpha_\uparrow \alpha_\downarrow^* \rho(\psi) \\ \alpha_\downarrow \alpha_\uparrow^* \rho(\psi) & p_\downarrow \rho(\psi) \end{bmatrix} = \begin{bmatrix} p_\uparrow^2 & p_\uparrow \alpha_\uparrow \alpha_\downarrow^* & \alpha_\uparrow \alpha_\downarrow^* p_\uparrow & \alpha_\uparrow \alpha_\downarrow^* \alpha_\uparrow \alpha_\downarrow^* \\ p_\uparrow \alpha_\downarrow \alpha_\uparrow^* & p_\uparrow p_\downarrow & \alpha_\uparrow \alpha_\downarrow^* \alpha_\downarrow \alpha_\uparrow^* & \alpha_\uparrow \alpha_\downarrow^* p_\downarrow \\ \alpha_\downarrow \alpha_\uparrow^* p_\uparrow & \alpha_\downarrow \alpha_\uparrow^* \alpha_\uparrow \alpha_\downarrow^* & p_\downarrow p_\uparrow & p_\downarrow \alpha_\uparrow \alpha_\downarrow^* \\ \alpha_\downarrow \alpha_\uparrow^* \alpha_\downarrow \alpha_\uparrow^* & \alpha_\downarrow \alpha_\uparrow^* p_\downarrow & p_\downarrow \alpha_\downarrow \alpha_\uparrow^* & p_\downarrow^2 \end{bmatrix}.$$

The qudits of F are $|\uparrow\rangle \otimes |\downarrow\rangle$ and $|\downarrow\rangle \otimes |\uparrow\rangle$ so the projection matrix to the subspace $[\text{qudit}(F)]$ of $\mathbb{C}^2 \otimes \mathbb{C}^2$ generated by those qudits is:

$$\begin{bmatrix} 0 & 0 & 0 & 0 \\ 0 & 1 & 0 & 0 \\ 0 & 0 & 1 & 0 \\ 0 & 0 & 0 & 0 \end{bmatrix}.$$

Hence the quantum logical entropy $h(F : \psi) = \operatorname{tr}\left[P_{[qudit(F)]}\rho(\psi) \otimes \rho(\psi)\right]$ is:

$$h\left(F:\psi\right) = \text{tr}\left(\begin{bmatrix} 0 & 0 & 0 & 0 \\ 0 & 1 & 0 & 0 \\ 0 & 0 & 1 & 0 \\ 0 & 0 & 0 & 0 \end{bmatrix}\begin{bmatrix} p_\uparrow^2 & p_\uparrow\alpha_\uparrow\alpha_\downarrow^* & \alpha_\uparrow\alpha_\downarrow^* p_\uparrow & \alpha_\uparrow\alpha_\downarrow^*\alpha_\uparrow\alpha_\downarrow^* \\ p_\uparrow\alpha_\downarrow\alpha_\uparrow^* & p_\uparrow p_\downarrow & \alpha_\uparrow\alpha_\downarrow^*\alpha_\downarrow\alpha_\uparrow^* & \alpha_\uparrow\alpha_\downarrow^* p_\downarrow \\ \alpha_\downarrow\alpha_\uparrow^* p_\uparrow & \alpha_\downarrow\alpha_\uparrow^*\alpha_\uparrow\alpha_\downarrow^* & p_\downarrow p_\uparrow & p_\downarrow\alpha_\uparrow\alpha_\downarrow^* \\ \alpha_\downarrow\alpha_\uparrow^*\alpha_\downarrow\alpha_\uparrow^* & \alpha_\downarrow\alpha_\uparrow^* p_\downarrow & p_\downarrow\alpha_\downarrow\alpha_\uparrow^* & p_\downarrow^2 \end{bmatrix}\right)$$

$$= \text{tr}\left(\begin{bmatrix} 0 & 0 & 0 & 0 \\ \alpha_\downarrow p_\uparrow\alpha_\uparrow^* & p_\uparrow p_\downarrow & \alpha_\uparrow\alpha_\downarrow\alpha_\uparrow^*\alpha_\downarrow^* & \alpha_\uparrow p_\downarrow\alpha_\downarrow^* \\ \alpha_\downarrow p_\uparrow\alpha_\uparrow^* & \alpha_\uparrow\alpha_\downarrow\alpha_\uparrow^*\alpha_\downarrow^* & p_\downarrow p_\uparrow & \alpha_\uparrow p_\downarrow\alpha_\downarrow^* \\ 0 & 0 & 0 & 0 \end{bmatrix}\right) = 2p_\uparrow p_\downarrow.$$

The second way to calculate the quantum logical entropy of the post-measurement state is using the Lüders mixture operation. The measurement of that spin-observable F goes from the pure state $\rho\left(\psi\right)$ to $P_\uparrow\rho\left(\psi\right)P_\uparrow + P_\downarrow\rho\left(\psi\right)P_\downarrow$. The calculation of the first term $P_\uparrow\rho\left(\psi\right)P_\uparrow$ can be usefully broken down into steps.

$$P_\uparrow\left(\rho\left(\psi\right)P_\uparrow\right) = \begin{bmatrix} 1 & 0 \\ 0 & 0 \end{bmatrix}\left(\begin{bmatrix} p_\uparrow & \alpha_\uparrow\alpha_\downarrow^* \\ \alpha_\downarrow\alpha_\uparrow^* & p_\downarrow \end{bmatrix}\begin{bmatrix} 1 & 0 \\ 0 & 0 \end{bmatrix}\right) = \begin{bmatrix} 1 & 0 \\ 0 & 0 \end{bmatrix}\begin{bmatrix} p_\uparrow & 0 \\ \alpha_\downarrow\alpha_\uparrow^* & 0 \end{bmatrix} = \begin{bmatrix} p_\uparrow & 0 \\ 0 & 0 \end{bmatrix}$$

and similarly

$$P_\downarrow\left(\rho\left(\psi\right)P_\downarrow\right) = \begin{bmatrix} 0 & 0 \\ 0 & 1 \end{bmatrix}\left(\begin{bmatrix} p_\uparrow & \alpha_\uparrow\alpha_\downarrow^* \\ \alpha_\downarrow\alpha_\uparrow^* & p_\downarrow \end{bmatrix}\begin{bmatrix} 0 & 0 \\ 0 & 1 \end{bmatrix}\right) = \begin{bmatrix} 0 & 0 \\ 0 & 1 \end{bmatrix}\begin{bmatrix} 0 & \alpha_\uparrow\alpha_\downarrow^* \\ 0 & p_\downarrow \end{bmatrix} = \begin{bmatrix} 0 & 0 \\ 0 & p_\downarrow \end{bmatrix}.$$

Hence the Lüders mixture operation yields:

$$P_\uparrow\rho\left(\psi\right)P_\uparrow + P_\downarrow\rho\left(\psi\right)P_\downarrow$$
$$= \begin{bmatrix} 1 & 0 \\ 0 & 0 \end{bmatrix}\begin{bmatrix} p_\uparrow & \alpha_\uparrow\alpha_\downarrow^* \\ \alpha_\downarrow\alpha_\uparrow^* & p_\downarrow \end{bmatrix}\begin{bmatrix} 1 & 0 \\ 0 & 0 \end{bmatrix} + \begin{bmatrix} 0 & 0 \\ 0 & 1 \end{bmatrix}\begin{bmatrix} p_\uparrow & \alpha_\uparrow\alpha_\downarrow^* \\ \alpha_\downarrow\alpha_\uparrow^* & p_\downarrow \end{bmatrix}\begin{bmatrix} 0 & 0 \\ 0 & 1 \end{bmatrix}$$
$$= \begin{bmatrix} p_\uparrow & 0 \\ 0 & p_\downarrow \end{bmatrix} = \hat{\rho}\left(\psi\right).$$

The logical entropy of the post-measurement density matrix $\hat{\rho}\left(\psi\right)$ is:

$$h\left(\hat{\rho}\left(\psi\right)\right) = 1 - \text{tr}\left[\hat{\rho}\left(\psi\right)^2\right] = 1 - p_\uparrow^2 - p_\downarrow^2 = 2p_\uparrow p_\downarrow = h\left(F:\psi\right)$$

since $1 = \left(p_\uparrow + p_\downarrow\right)^2 = p_\uparrow^2 + p_\downarrow^2 + 2p_\uparrow p_\downarrow$.

The third way to calculate the quantum logical entropy of $\hat{\rho}\left(\psi\right)$ is to sum the absolute squares of the non-zero off-diagonal terms in the pure state density matrix $\rho\left(\psi\right)$ (so $h\left(\rho\left(\psi\right)\right) = 0$) that are zeroed in the transition to the post-measurement density matrix $\hat{\rho}\left(\psi\right)$, i.e.,

$$\rho\left(\psi\right) = \begin{bmatrix} p_\uparrow & \alpha_\uparrow\alpha_\downarrow^* \\ \alpha_\downarrow\alpha_\uparrow^* & p_\downarrow \end{bmatrix} \rightsquigarrow \begin{bmatrix} p_\uparrow & 0 \\ 0 & p_\downarrow \end{bmatrix} = \hat{\rho}\left(\psi\right),$$

and that sum is:

$$2\alpha_\uparrow \alpha_\downarrow^* \alpha_\downarrow \alpha_\uparrow^* = 2p_\uparrow p_\downarrow = h\left(\hat{\rho}\left(\psi\right)\right) - h\left(\rho\left(\psi\right)\right) = h\left(F:\psi\right).$$

The Tables 4.2 and 4.3 show how the definitions of quantum logical entropy are the linearized versions of the definitions of logical entropy of set partitions. Those definitions by themselves prove nothing concerning our thesis about QM mathematics. But the usual treatment of density matrices and the Lüders mixture operation representing projective measurement does not involve classical or quantum logical entropy which is little known in the standard texts that use the quantum version of Shannon entropy, i.e., von Neumann entropy. Hence the close and precise connection (down to the level: indistinctions $\sim$ coherences) between the mathematics of partitions and projective measurement in terms of density matrices was not part of the standard presentation of QM mathematics which only uses von Neumann entropy. This is a case where the relevant QM mathematics (about quantum logical entropy) had to be developed to show how that QM mathematics (e.g., Tables 4.2 and 4.3 and the Measuring Measurement Theorem) is the vector space version of the mathematics of partitions.

Quantum measurement creates distinctions, e.g., the distinction between spin-up and spin-down in the spin example, and quantum logical entropy precisely measures those distinctions. We started 'classically' with a universe set of distinct elements

The set can be thought of as being originally fully distinct, while each partition collects together blocks whose distinctions are factored out. Each block represents elements that are associated with an equivalence relation on the set. Then, the elements of a block are indistinct among themselves while different blocks are distinct from each other, given an equivalence relation. (Tamir et al. 2022, p. 1)

Then the Yoga of Linearization translated the concepts of information-as-distinctions to the corresponding vector space concept–which for Hilbert spaces, would be the concepts of quantum information-as-qudits.

With these concepts in mind, it seems that the extension of this framework of partitions and distinctions to the study of quantum systems may bring new insights into problems of quantum state discrimination, quantum cryptography, and quantum channel capacity. In fact, in these problems, we are, in one way or another, interested in a distance measure between distinguishable states, which is exactly the kind of knowledge the logical entropy is associated with. (Tamir et al. 2022, p. 1)

4.5 Group Representations

4.5.1 The Yoga Extended to Group Representations

Group representation theory is an important part of the mathematics of QM (Weyl 1950; French 1999, 2013). This immediately supports our thesis since a group representation is essentially a 'dynamic' or 'active' way to define an equivalence relation; each group operation transforms an element into an equivalent or symmetric element.

Given a *set* G indexing (associative) mappings $\left\{R_g : U \to U\right\}_{g \in G}$ on a set U, what are the conditions on the set of mappings so that it is a set representation of a group? Define the binary relation R on $U \times U$ by:

$$\left(u_i, u_k'\right) \in R \text{ if } \exists g \in G \text{ such that } R_g\left(u_i\right) = u_k'.$$

Then the conditions that make R_g into a group representation are the conditions that imply R is an equivalence relation:

1. existence of the identity $1_U \in G$ implies reflexivity of R;
2. existence of inverses implies symmetry of R; and
3. closure under products, i.e., for $g, g' \in G$, $\exists g'' \in G$ such that $R_{g''} = R_{g'} R_g$, implies transitivity of R.

Thus a set representation of a group G (or group action on a set) is essentially a 'dynamic' way to define an equivalence relation on the set (Weyl 1949, p. 73; Castellani 2003). The ordered pairs (u_i, u_k) where $\exists g \in G$ such that $R_g\left(u_i\right) = u_k$ are the indistinctions of a partition on U called the *orbit partition*. A subset $S \in U$ is *invariant* if for any $u_i \in S$ and any $g \in G$, $R_g\left(u_i\right) \in S$. The minimal invariant (or irreducible) subsets of the set representation are the *orbits*, and they are the equivalence classes of the equivalence relation or blocks of the orbit partition. The restriction of a set representation of a group to an orbit is an *irreducible representation* or *irrep*.

Given a numerical attribute $f : U \to \mathbb{R}$, let $\tilde{f} : \mathbb{R}^n \to \mathbb{R}^n$ be the $n \times n$ diagonal matrix with the diagonal entries $\tilde{f}_{ii} = f\left(u_i\right)$. Let $M : \mathbb{R}^n \to \mathbb{R}^n$ be an $n \times n$ matrix that commutes with $\tilde{f}$ in the sense that the following diagram commutes:

$$
\begin{array}{ccc}
\mathbb{R}^n & \xrightarrow{\ M\ } & \mathbb{R}^n \\
\tilde{f} \downarrow & & \downarrow \tilde{f}, \\
\mathbb{R}^n & \xrightarrow{\ M\ } & \mathbb{R}^n
\end{array}
$$

i.e., $\tilde{f} M = M \tilde{f}$. Then computing the ik-entry in the two ways gives: $\left(\tilde{f} M\right)_{ik} = f\left(u_i\right) M_{ik} = M_{ik} f\left(u_k\right) = \left(M \tilde{f}\right)_{ik}$ so $\left(f\left(u_i\right) - f\left(u_k\right)\right) M_{ik} = 0$. Hence if (u_i, u_k) is a dit of the partition f^{-1}, i.e., $f\left(u_i\right) \neq f\left(u_k\right)$, then $M_{ik} = 0$.

The vector space version of this result is that for two commuting (diagonalizable) operators $FG = GF$, then when both are represented in an eigen basis of F, then if (u_i, u_k) is a qudit of F, i.e., u_i and u_k have different eigenvalues, then $G_{ik} = 0$ (Tinkham 1964, p. 4).

A numerical attribute $f : U \to \mathbb{R}$ is said to *commute* with the set representation $R = \{R_g\}_{g \in G}$ if for any R_g, the following diagram commutes:

$$
\begin{array}{ccc}
U & \xrightarrow{R_g} & U \\
 & \searrow^{f} \quad \downarrow^{f} & \\
 & \mathbb{R} &
\end{array}
$$

The blocks $f^{-1}(r)$ for $r \in f(U)$ for a commuting f are invariant subsets of U under R. Thus the partition f^{-1} for a commuting f is refined by the orbit partition since orbits are the *minimal* invariant subsets. A set of commuting attributes $f, g,..., h$ is said to be *complete* (a CSCA) if the join $f^{-1} \vee g^{-1} \vee ... \vee h^{-1}$ is the orbit partition.

Given a finite group G and a finite-dimensional vector space V over $\mathbb{C}$, a *vector space representation* of the group G is a map $R : G \to GL(V)$ where $g \longmapsto R_g : V \to V$ from G to invertible linear transformations on V such that $R_1 = I$ and $R_{g'} R_g = R_{g'g}$. Using the Yoga to lift set concepts to vector space concepts, the notion of a minimal invariant subset or orbit yields the notion of a minimal invariant subspace which is called an *irreducible subspace* of V. Just as the orbits of a set representation form a partition of U, so the irreducible subspaces of a vector space representation form a direct-sum decomposition of V. The restriction of a vector space representation of a group to an irreducible subspace is an *irreducible representation* or *irrep*.

To apply the Yoga, let $\{R_g\}_{g \in G}$ be a set representation of G on a set U that is an ON basis set for V and let $f : U \to \mathbb{R}$ be a commuting attribute. Then for any $u_i \in U$ and any $g \in G$, $R_g(u_i) = u_k \in U$ so, by commutativity, $f(u_i) = f(R_g(u_i)) = f(u_k)$. That is, for any u_i with an f-value $f(u_i)$, any R_g in the representation takes u_i to another element u_k with the same f-value. That means the orbit partition for the representation refines the inverse-image partition defined by f. Thus the commuting attribute f is constant on the orbits of the orbit partition of the representation.

Then an observable operator $F : V \to V$ is said to *commute* with the vector space representation if the following diagram commutes for every g in the group.

$$
\begin{array}{ccc}
V & \xrightarrow{R_g} & V \\
F \downarrow & & \downarrow F \\
V & \xrightarrow{R_g} & V
\end{array}
$$

Starting with a basis element u_i (in an ON basis U for F) and going around the square counter-clockwise, yields $Fu_i = f(u_i)u_i$ where $f : U \to \mathbb{R}$ is the eigenvalue function for F. And then applying R_g gives $R_g(f(u_i)u_i) = f(u_i)R_g(u_i)$. Going around the square clockwise gives $F(R_g(u_i))$. And the commutativity gives:

Table 4.4 Summary of Yoga of linearization for group representations

Yoga	Set representations	Vector space representations
Representation	$\{R_g : U \to U\}_{g \in G}$	$\{R_g : V \to V\}_{g \in G}$
Min. Invariants	Orbits	Irreducible subspaces
Partition	Orbit partition	DSD of irred. subspaces
Irreps	Representation on orbits	Rep. on irred. subspaces
Commuting	$f : U \to \mathbb{R}, \forall g, f R_g = f$	$F : V \to V, \forall g, F R_g = R_g F$
Invariants	$f^{-1}(r)$ commuting f	Eigenspaces commuting F
Complete set	CSCA	CSCO
Schur's Lemma	Comm. $f \upharpoonright$ orbit constant	Comm. $F \upharpoonright$ irred. space constant

$f(u_i) R_g(u_i) = F(R_g(u_i))$ which means that R_g carries u_i to a vector with the same f-value, i.e., same eigenvalue. Hence the set-concept of a commuting f extends by the Yoga to the usual concept of an observable F commuting with a vector space representation.

The eigenspaces of a commuting F are invariant under R. Then the DSD of eigenspaces for a commuting F is refined (defined in the obvious way) by the DSD of irreducible subspaces. A set of commuting operators F, G,..., H is said to be *complete* (a CSCO) if the join-like operation on their DSDs has all its subspaces as irreducible–which is the vector space version of a CSCA for set representations.

Schur's Lemma (set case): A commuting f restricted to (i.e., $f \upharpoonright$) an irreducible subset (i.e., an orbit) is a constant function.

Schur's Lemma (vector space case): A commuting F restricted to (i.e., $F \upharpoonright$) an irreducible subspace is a constant operator.

The Yoga of Linearization applied to group representations is illustrated in Table 4.4.

4.5.2 Examples of Set and Vector-Space Representations

Our theme is showing how the mathematics of QM, in this case the mathematics of vector space representations, is the vector space version of partitional mathematics on sets. We focus first on the relatively simple task of giving examples of the set versions.

Example 1: Let $U = \{0, 1, 2, 3, 4, 5\}$ and let $G = S_2 = \{1, \sigma\}$ (symmetric group on two elements) where $R_1 = 1_U$ and $R_\sigma(u) = u + 3 \mod 6$ as shown in Fig. 4.9.

There are 3 orbits: $\{0, 3\}$, $\{1, 4\}$, and $\{2, 5\}$, and they form the orbit partition of U. Those three orbits are the minimal invariant subsets, the symmetry-adapted eigen-alternatives defined by the symmetry group's S_2 action on U. The "irreducible representations" or irreps in the set case are just the restrictions of the representation to the orbits, e.g., $R \upharpoonright \{0, 3\}$ where R_σ just permutes 0 and 3.

Fig. 4.9 Set representation of S_2 on U

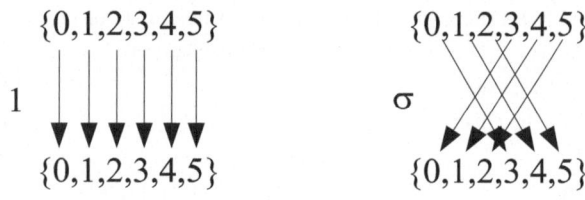

For Example 1, consider the attribute $f : U = \{0, 1, 2, 3, 4, 5\} \rightarrow \{0, 1, 2\}$ where $f(n) = n \bmod 3$. This attribute commutes with the set representation of S_2, namely $R_1 = 1_U$ and $R_\sigma(u) = u + 3 \bmod 6$, and accordingly, by Schur's Lemma (set version), the attribute is constant on each orbit $\{0, 3\}$, $\{1, 4\}$, and $\{2, 5\}$. In this case, the blocks of the inverse-image partition $\{f^{-1}(0), f^{-1}(1), f^{-1}(2)\}$ equal the blocks of the orbit partition, so this attribute by itself is a complete set of attributes in that its eigenvalues 0, 1, 2 suffice to characterize the orbits.

The set-version of the superposition principle in QM is to consider the blocks in a partition, such as the orbit partition, as indefinite superpositions of the elements in the blocks. Moving to the representation of a subgroup (breaking symmetry) makes less indistinctions and thus more distinctions so it moves from the indefinite blocks in the original orbit partition to the more definite blocks in the orbit partition of the subgroup representation. In the example, this is trivial since the only subgroup of S_2 is the identity subgroup whose orbits are the singletons of U, so the orbit partition of the identity subgroup is the discrete partition 1_U.

Example 2: Let $U = \{0, 1, ..., 11\}$ where $S_2 = \{1, \sigma\}$ is represented by the operations $R_1 = 1_U$ and $R_\sigma(n) = n + 6 \bmod (12)$. The orbits are $\{0, 6\}, \{1, 7\}, \{2, 8\}$, $\{3, 9\}, \{4, 10\}$, and $\{5, 11\}$. Consider the attribute $f : U \rightarrow \{0, 1\}$ where $f(n) = n \bmod (2)$. This attribute commutes with the symmetry group and is thus constant on the orbits. Now the blocks in the inverse-image partition are larger than the orbits, i.e., $f^{-1}(0) = \{0, 2, 4, 6, 8, 10\}$ and $f^{-1}(1) = \{1, 3, 5, 7, 9, 11\}$ so the orbit partition strictly refines $f^{-1} = \{f^{-1}(r)\}_{r \in f(U)}$. Thus this attribute corresponds to a degenerate measurement in that the two eigenvalues do not suffice to characterize the orbits.

Consider another attribute $g : U \rightarrow \{0, 1, 2\}$ where $g(n) = n \bmod (3)$. This attribute commutes with the symmetry group and is thus constant on the orbits by Schur's Lemma. The blocks in the inverse-image partition are: $g^{-1}(0) = \{0, 3, 6, 9\}$, $g^{-1}(1) = \{1, 4, 7, 10\}$, and $g^{-1}(2) = \{2, 5, 8, 11\}$. The blocks in the join of the two partitions $\{f^{-1}(r)\}$ and $\{g^{-1}(s)\}$ are the non-empty intersections of the blocks as shown in Table 4.5.

Thus f and g form a Complete Set of Commuting Attributes (CSCA) to characterize the eigen-alternatives, the orbits, by the kets of ordered pairs of their eigenvalues given in the last column of Table 4.5.

A *vector space representation* of a group G is given by an assignment of an invertible linear operator $R_g : V \rightarrow V$ to each $g \in G$ such that $R_1 = 1_V$ and $R_{g'} R_g = R_{g'g}$.

Table 4.5 f and g as a complete set of commuting attributes

$f^{-1}(r)$	$g^{-1}(s)$	$f^{-1}(r) \cap g^{-1}(s)$	$|r, s\rangle$
$\{0, 2, 4, 6, 8, 10\}$	$\{0, 3, 6, 9\}$	$\{0, 6\}$	$|0, 0\rangle$
$\{0, 2, 4, 6, 8, 10\}$	$\{1, 4, 7, 10\}$	$\{4, 10\}$	$|0, 1\rangle$
$\{0, 2, 4, 6, 8, 10\}$	$\{2, 5, 8, 11\}$	$\{2, 8\}$	$|0, 2\rangle$
$\{1, 3, 5, 7, 9, 11\}$	$\{0, 3, 6, 9\}$	$\{3, 9\}$	$|1, 0\rangle$
$\{1, 3, 5, 7, 9, 11\}$	$\{1, 4, 7, 10\}$	$\{1, 7\}$	$|1, 1\rangle$
$\{1, 3, 5, 7, 9, 11\}$	$\{2, 5, 8, 11\}$	$\{5, 11\}$	$|1, 2\rangle$

Example 3: The multiplicative group $S_2 \times S_2$ written additively is the Klein four-group $G = \mathbb{Z}_2 \times \mathbb{Z}_2 = \{(0, 0), (1, 0), (0, 1), (1, 1)\}$. The complex vector space $\{\mathbb{Z}_2 \times \mathbb{Z}_2 \to \mathbb{C}\}$ of all complex-valued functions on the four-element set $\mathbb{Z}_2 \times \mathbb{Z}_2$ is the *Cayley group space* of that group. A basis for the four-dimensional space $\mathbb{C}^4$ is the set of functions $|g'\rangle$ which take value 1 on g' and 0 on the other $g \in G$. Then the action of the group on this space is defined by $R_g(|g'\rangle) = |g + g'\rangle$ (or $|gg'\rangle$ if the group operation was written multiplicatively). Thus the group action just permutes the basis vectors in the Cayley group space and would be represented by permutation matrices. The non-identity operators have the matrices;

$$R_{(1,0)} = \begin{bmatrix} 0 & 1 & 0 & 0 \\ 1 & 0 & 0 & 0 \\ 0 & 0 & 0 & 1 \\ 0 & 0 & 1 & 0 \end{bmatrix} \begin{matrix} (0,0) \\ (1,0) \\ (0,1) \\ (1,1) \end{matrix} ; R_{(0,1)} = \begin{bmatrix} 0 & 0 & 1 & 0 \\ 0 & 0 & 0 & 1 \\ 1 & 0 & 0 & 0 \\ 0 & 1 & 0 & 0 \end{bmatrix} ; R_{(1,1)} = \begin{bmatrix} 0 & 0 & 0 & 1 \\ 0 & 0 & 1 & 0 \\ 0 & 1 & 0 & 0 \\ 1 & 0 & 0 & 0 \end{bmatrix}.$$

Since the group is Abelian, each of these operators can be viewed as an observable that commutes with the R_g for $g \in G$, so its eigenspaces will be invariant under the group operations.

For $R_{(1,0)}$, the invariant eigenspaces with their eigenvalues and generating eigenvectors are:

$$\left\{ \begin{bmatrix} 1 \\ -1 \\ 1 \\ -1 \end{bmatrix}, \begin{bmatrix} 1 \\ -1 \\ -1 \\ 1 \end{bmatrix} \right\} \leftrightarrow \lambda = -1, \left\{ \begin{bmatrix} 1 \\ 1 \\ 1 \\ 1 \end{bmatrix}, \begin{bmatrix} 1 \\ 1 \\ -1 \\ -1 \end{bmatrix} \right\} \leftrightarrow \lambda = 1.$$

For $R_{(0,1)}$, we have:

$$\left\{ \begin{bmatrix} 1 \\ -1 \\ -1 \\ 1 \end{bmatrix}, \begin{bmatrix} 1 \\ 1 \\ -1 \\ -1 \end{bmatrix} \right\} \leftrightarrow \lambda = -1, \left\{ \begin{bmatrix} 1 \\ 1 \\ 1 \\ 1 \end{bmatrix}, \begin{bmatrix} 1 \\ -1 \\ 1 \\ -1 \end{bmatrix} \right\} \leftrightarrow \lambda = 1.$$

Since these two operators commute, their eigenspace DSDs commute so we can take their join. The blocks of the join are a DSD and are automatically invariant. Since the blocks of the join are one-dimensional, those four subspaces are also irreducible and thus those two operators form a complete set of commuting operators (CSCO). The commuting operators always have a set of simultaneous eigenvectors, and we have arranged the generating eigenvectors of the eigenspaces so that they are all simultaneous eigenvectors which can, as usual, be characterized by kets using the respective eigenvalues;

$$
\begin{bmatrix} 1 \\ 1 \\ 1 \\ 1 \end{bmatrix} = |1, 1\rangle ; \quad
\begin{bmatrix} 1 \\ -1 \\ 1 \\ -1 \end{bmatrix} = |-1, 1\rangle ; \quad
\begin{bmatrix} 1 \\ 1 \\ -1 \\ -1 \end{bmatrix} = |1, -1\rangle ; \quad
\begin{bmatrix} 1 \\ -1 \\ -1 \\ 1 \end{bmatrix} = |-1, -1\rangle .
$$

Restricting the group representation to these four irreducible subspaces gives the four irreducible representations or irreps of the group. Since any vector can be uniquely decomposed into the sum of vectors in the irreducible subspaces, the representation on the whole space can be expressed, in the obvious sense, as the direct sum of the irreps.

In the set case, moving to smaller invariant subsets by making distinctions gives more distinct elements, so in the vector space case, moving to smaller invariant subspaces give more distinct alternatives. The minimal invariant subspaces, i.e., the irreducible subspaces, thus give the maximally-distinct invariant subspaces, and the representation restricted to those subspaces gives the maximally-distinct symmetry-respecting alternatives, i.e., the irreps.

This can be shown by giving a geometric version of the representation. Consider a rectangle $\begin{smallmatrix} a & b \\ d & c \end{smallmatrix}$ under the operations of flipping on the horizontal axis $\sigma_h : \begin{smallmatrix} a & b \\ d & c \end{smallmatrix} \longmapsto \begin{smallmatrix} d & c \\ a & b \end{smallmatrix}$ and flipping on the vertical axis: $\sigma_v : \begin{smallmatrix} a & b \\ d & c \end{smallmatrix} \longmapsto \begin{smallmatrix} b & a \\ c & d \end{smallmatrix}$ as well as their composition $\sigma_{hv} : \begin{smallmatrix} a & b \\ d & c \end{smallmatrix} \longmapsto \begin{smallmatrix} c & d \\ b & a \end{smallmatrix}$ and the identity. This gives the same $S_2 \times S_2$ group with the multiplication Table 4.6.

Table 4.6 Multiplication table for $S_2 \times S_2$ as symmetry group resulting from flipping on the horizontal and vertical axes

2nd \ 1st	1	σ_h	σ_v	σ_{hv}
1	$\begin{smallmatrix} a & b \\ d & c \end{smallmatrix}$	$\begin{smallmatrix} d & c \\ a & b \end{smallmatrix}$	$\begin{smallmatrix} b & a \\ c & d \end{smallmatrix}$	$\begin{smallmatrix} c & d \\ b & a \end{smallmatrix}$
σ_h	$\begin{smallmatrix} d & c \\ a & b \end{smallmatrix}$	$\begin{smallmatrix} a & b \\ d & c \end{smallmatrix}$	$\begin{smallmatrix} c & d \\ b & a \end{smallmatrix}$	$\begin{smallmatrix} b & a \\ c & d \end{smallmatrix}$
σ_v	$\begin{smallmatrix} b & a \\ c & d \end{smallmatrix}$	$\begin{smallmatrix} c & d \\ b & a \end{smallmatrix}$	$\begin{smallmatrix} a & b \\ d & c \end{smallmatrix}$	$\begin{smallmatrix} d & c \\ a & b \end{smallmatrix}$
σ_{hv}	$\begin{smallmatrix} c & d \\ b & a \end{smallmatrix}$	$\begin{smallmatrix} b & a \\ c & d \end{smallmatrix}$	$\begin{smallmatrix} d & c \\ a & b \end{smallmatrix}$	$\begin{smallmatrix} a & b \\ d & c \end{smallmatrix}$

Table 4.7 Character table for $S_2 \times S_2$

	$\begin{smallmatrix} a\ b \\ d\ c \end{smallmatrix}$	$\begin{smallmatrix} d\ c \\ a\ b \end{smallmatrix}$	$\begin{smallmatrix} b\ a \\ c\ d \end{smallmatrix}$	$\begin{smallmatrix} c\ d \\ b\ a \end{smallmatrix}$
χ_1	1	1	1	1
χ_2	1	-1	1	-1
χ_3	1	1	-1	-1
χ_4	1	-1	-1	1

Then the basis vectors for the four irreducible subspaces of the Cayley space (sometimes called the *irreducible basis vectors*) are given in the character Table 4.7.

The one-dimensional subspace generated by any one of the vectors is invariant under the group operations. For instance, if we apply σ_v to χ_3, we get:

$$
R_{\sigma_v}(\chi_3) = R_{\sigma_v}\left(\boxed{\begin{smallmatrix} a\ b \\ d\ c \end{smallmatrix}} + \boxed{\begin{smallmatrix} d\ c \\ a\ b \end{smallmatrix}} - \boxed{\begin{smallmatrix} b\ a \\ c\ d \end{smallmatrix}} - \boxed{\begin{smallmatrix} c\ d \\ b\ a \end{smallmatrix}} \right)
$$

$$
= \boxed{\begin{smallmatrix} b\ a \\ c\ d \end{smallmatrix}} + \boxed{\begin{smallmatrix} c\ d \\ b\ a \end{smallmatrix}} - \boxed{\begin{smallmatrix} a\ b \\ d\ c \end{smallmatrix}} - \boxed{\begin{smallmatrix} d\ c \\ a\ b \end{smallmatrix}} = -\chi_3.
$$

These four irrep basis vectors represent the maximally distinct (e.g., mutually orthogonal) eigen-forms that respect the symmetry operations. Each one is a super-position of the original four basis vectors which were not "symmetry-adapted." By superposing the original basis vectors in this way, we get the maximally distinct eigen-forms obeying the symmetry group.

This example also illustrates how the vector space representations, as opposed to the set representations, generate more variety. Instead of the Cayley space, we consider the *Cayley set* which is just the set G of group operations which we might represent by the set of four configurations obtained from the initial configuration $\boxed{\begin{smallmatrix} a\ b \\ d\ c \end{smallmatrix}}$, namely $\left\{ \boxed{\begin{smallmatrix} a\ b \\ d\ c \end{smallmatrix}}, \boxed{\begin{smallmatrix} d\ c \\ a\ b \end{smallmatrix}}, \boxed{\begin{smallmatrix} b\ a \\ c\ d \end{smallmatrix}}, \boxed{\begin{smallmatrix} c\ d \\ b\ a \end{smallmatrix}} \right\}$. Each group operation such as R_{σ_v} acts on these four elements as indicated by the row in the multiplication table. Thus we have a set representation of the group whose orbits will be the minimal invariant subsets. But in every Cayley set representation, there is only one orbit since any element can be mapped to any other element by the action of one of the four operators. That is, the Cayley group action is said to be *transitive* in the sense that if for any $u_i, u_k \in U$, $\exists g \in G$ such that $R_g(u_i) = u_k$. A transitive set representation has only one orbit, all of U. Any set "irreducible representation" is transitive. Thus the only eigen-form we get from the Cayley set representation of $S_2 \times S_2$ is the minimal invariant subset or orbit $\left\{ \boxed{\begin{smallmatrix} a\ b \\ d\ c \end{smallmatrix}}, \boxed{\begin{smallmatrix} d\ c \\ a\ b \end{smallmatrix}}, \boxed{\begin{smallmatrix} b\ a \\ c\ d \end{smallmatrix}}, \boxed{\begin{smallmatrix} c\ d \\ b\ a \end{smallmatrix}} \right\}$ which corresponds to χ_1 in the Cayley vector space representation. By working with a vector space representation over $\mathbb{C}$, we immediately get a richer set of irreps.

When the group being represented is non-Abelian, not to mention continuous, the mathematical difficulty greatly increases, but our theme remains the same. Our point is that the key concepts for group representations in QM come out of the partitional mathematics of distinctions and indistinctions. The irreps, in both the set and vector space cases, give all the different ways that indefinite states are defined consistent with the indistinction-creating symmetries of the group. Those minimal invariant or irreducible subspaces can be defined using the join of commuting DSDs. The irreps fill out the symmetry-adapted possibilities.

Since the group action creates indistinctions as symmetries, moving to the action of a subgroup means less indistinctions and more distinctions, i.e., symmetry-breaking. Thus symmetry-breaking is the second method, in addition to the join of DSDs, to move from an indefinite state to a more definite state. Making all possible distinctions means moving to the identity subgroup where the orbit partition is the discrete partition representing classical reality.

The indefinite is rendered definite by the distinctions made by the joins of partitions collected together in a complete set. In quantum mechanics, the definite eigenstates of the observables are carved out by the joins of vector space partitions of Dirac's CSCOs. In vector space representations, the properties of the eigen-alternatives are determined by the irreps of the symmetry group of the Hamiltonian or as the elementary particles themselves are determined by the irreps of groups in particle physics.

In Heisenberg's philosophical terms, the irreducible representations of certain symmetry groups of particle physics determine the fundamental eigen-forms that the substance (energy) can take.

> The elementary particles are therefore the fundamental forms that the substance energy must take in order to become matter, and these basic forms must in some way be determined by a fundamental law expressible in mathematical terms. ... The real conceptual core of the fundamental law must, however, be formed by the mathematical properties of the symmetry it represents. (Heisenberg 1961, pp. 16–7)

For a certain symmetry group of particle physics, "an elementary particle 'is' an irreducible unitary representation of the group" (Sternberg 1994, p. 149). Thus our partitional approach comports with "the soundness of programs that ground particle properties in the irreducible representations of symmetry transformations..." (Fine 1986, p. 171). These alternatives are carved out by the joins of the vector space partitions of CSCOs–which constitute a "systematic theory ... established for the rep group based on Dirac's CSCO (complete set of commuting operators) approach in quantum mechanics" (Chen et al. 1985, p. 211, also 2002; Wang 2004).

This all goes back to the transformations of group representations being 'dynamic' or 'active' ways to define partitions (equivalence relations) and their vector space versions (DSDs).

References

Auletta G (2019) The quantum mechanics conundrum: interpretations and foundations. Springer Nature, Cham, Switzerland

Auletta G, Fortunato M, Parisi G (2009) Quantum mechanics. Cambridge University Press, Cambridge, UK

Bennett CH (2003) Quantum information: qubits and quantum error correction. Int J Theor Phys 42:153–176. https://doi.org/10.1023/A:1024439131297

Bertlmann RA, Krammer P (2008) Bloch vectors for qudits. J Phys A: Math Theor 41:235303. https://doi.org/10.1088/1751-8113/41/23/235303

Brukner Č, Zeilinger A (2003) Information and fundamental elements of the structure of quantum theory. In: Castell L, Ischebeck O (eds) Time, quantum and information. Springer, Berlin, pp 323–354

Castellani E (2003) Symmetry and equivalence. In: Brading K, Castellani E (eds) Symmetries in physics: philosophical reflections. Cambridge University Press, Cambridge, pp 425–436

Chen J-Q, Gao M-J, Ma G-Q (1985) The representation group and its application to space groups. Rev Mod Phys 57:211–278

Chen J-Q, Ping J, Wang F (2002) Group representation theory for physicists, 2nd ed. World Scientific, Singapore

Dirac PAM (1958) The principles of quantum mechanics, 4th ed. Clarendon Press, Oxford

Eddington AS (1947) New pathways in science (Messenger Lectures 1934). Cambridge University Press, Cambridge, UK

Ellerman D (2009) Counting distinctions: on the conceptual foundations of Shannon's information theory. Synthese 168:119–149. https://doi.org/10.1007/s11229-008-9333-7

Ellerman D (2021) New foundations for information theory: logical entropy and Shannon entropy. Springer Nature, Cham, Switzerland

Ellerman D (2022) Introduction to logical entropy and its relationship to Shannon entropy. 4Open Special Issue: Logical Entropy 5:1–33. https://doi.org/10.1051/fopen/2021004

Fano U (1957) Description of states in quantum mechanics by density matrix and operator techniques. Rev. Mod. Phys. 29:74–93

Feller W (1968) An introduction to probability theory and its applications, vol 1, 3rd ed. Wiley, New York

Feynman RP (1951) The concept of probability in quantum mechanics. In: Second Berkeley symposium on mathematical statistics and probability. University of California Press, pp 533–541

Feynman RP, Leighton RB, Sands M (1965) The Feynman lectures on physics: quantum mechanics, vol III. Addison-Wesley, Reading, MA

Feynman RP, Hibbs AR (1965) Quantum Mechanics and Path Integrals. McGraw-Hill, New York

Fine A (1986) The Shaky Game: Einstein, Realism, and the quantum theory. University of Chicago Press, Chicago

French S (1999) Models and mathematics in physics: the role of group theory. In: Butterfield J, Pagonis C (eds) From physics to philosophy. Cambridge University Press, Cambridge, UK, pp 187–207

French S (2013) Whither wave function realism? In: Ney A, Albert DZ (eds) The wave function: essays on the metaphysics of quantum mechanics. Oxford University Press, New York, pp 76–90

Heisenberg W (1961) Planck's discovery and the philosophical problems of atomic physics. On modern physics. Clarkson N. Potter Inc, New York, pp 3–19

Hoffman K, Kunze R (1961) Linear algebra. Prentice-Hall, Englewood Cliffs, NJ

Hughes RIG (1989) The structure and interpretation of quantum mechanics. Harvard University Press, Cambridge

Jaeger G (2007) Quantum information: an overview. Springer Science+Business Media, New York

Kolmogorov AN (1983) Combinatorial foundations of information theory and the calculus of probabilities. Russian Math Surv 38:29–40

Leibniz GW (1996) New essays on human understanding. Cambridge University Press, Cambridge, UK

Maudlin T (2019) Philosophy of physics: quantum theory. Princeton University Press, Princeton, NJ

Nielsen M, Chuang I (2000) Quantum computation and quantum information. Cambridge University Press, Cambridge

Norsen T (2017) Foundations of quantum mechanics. Springer International, Cham, Switzerland

Sakurai JJ, Napolitano J (2011) Modern quantum mechanics, 2nd ed. Addison-Wesley, Boston
Stachel J (1986) Do quanta need a new logic? In: Colodny RG (ed) From quarks to quasars: philosophical problems of modern physics. University of Pittsburgh Press, Pittsburgh, pp 229–347
Sternberg S (1994) Group theory and physics. Cambridge University Press, Cambridge
Stone MH (1932) On one-parameter unitary groups in Hilbert Space. Ann Math 33:643–648
Tamir B, Piava IL, Schwartzman-Nowik Z, Cohen E (2022) Quantum logical entropy: fundamentals and general properties. 4Open Special Issue: Logical Entropy 5:1–14. https://doi.org/10.1051/fopen/2021005
Tinkham M (1964) Group theory and quantum mechanics. Dover, Mineola, NY
Wang F (2004) A conceptual review of the new approach to group representation theory. In: Feng DH, Iachello F, Ping JL, Wang F (eds) The beauty of mathematics in science: the intellectual path of JQ Chen. World Scientific, Singapore, pp 1–8
Weyl H (1949) Philosophy of mathematics and natural science. Princeton University Press, Princeton
Weyl H (1950) The theory of groups and quantum mechanics. Dover Publications, New York
Wigner EP (1967) The unreasonable effectiveness of mathematics in the natural sciences. Symmetries and reflections. Indiana University Press, Bloomington, IN, pp 222–237
Wilson RJ (1972) Introduction to Graph Theory. Longman, London
Zeilinger A (1999) Experiment and the foundations of quantum physics. In: Bederson B (ed) More things in heaven and earth. Springer, Berlin, pp 482–498
Zurek WH (2003) Decoherence, einselection, and the quantum origins of the classical. Rev Mod Phys 75:715–775

Chapter 5
Conclusions

*The combination of indefiniteness of value with definite
probabilities of possible outcomes can be compactly referred to
as potentiality, a term suggested by Heisenberg (1962). When a
physical variable which initially is merely potential acquires a
definite value, it can be said to be actualized. So far, the only
processes we have mentioned in which potentialities are
actualized are measurements, but in a non-anthropocentric view
of physical theory the measurement process is only a special
case of the interaction of systems, of special interest to scientists
because knowledge is thereby obtained, but not fundamental
from the standpoint of physical reality itself.*

Abner Shimony (1993)

Abstract This chapter briefly recapitulates the main points of the Objective Indefiniteness Interpretation of QM without the detailed development and examples. To the non-philosophical quantum theorist, there is a simple litmus test. Does a superposition state objectively have a definite or indefinite value prior to measurement? If the answer is "objectively indefinite," then the quantum theorist is a (perhaps "closet") supporter of the Objectively Indefinite Interpretation.

Keywords Quantum states · Quantum observables · Projective measurement · Commuting observables · Schrödinger equation · Particle indistinguishability · Lüders mixture operation · Group representations

5.1 Overview

Where does the mathematical machinery of quantum mathematics (not the physics[1]) come from? Surely an answer to that question will contribute to a better understanding of the key analytical concepts and a better comprehension of the underlying physical reality that is so different from the fully-definite intuitive world of classical physics.

---

[1] The focus on the mathematics is evidenced by Planck's constant not being used in the entire book.

© The Author(s), under exclusive license to Springer Nature Switzerland AG 2024   105
D. Ellerman, *Partitions, Objective Indefiniteness, and Quantum Reality*,
SpringerBriefs in Philosophy, https://doi.org/10.1007/978-3-031-61786-7_5

We have argued that the mathematics of QM comes from the vector (particularly, Hilbert) space version of the mathematics of partitions (or equivalence relations). The Yoga of Linearization is part of the mathematical folklore and has been previously used to help explain QM, e.g., by Hermann Weyl. Partitions are the set-level logical concepts to formalize distinctions and indistinctions which thus shows that distinguishability and indistinguishability are key analytical concepts in QM. In this and previous works (Ellerman 2021, 2024), it is argued that the basic notion of information is information-as-distinctions or information-as-distinguishability. Since those are the key analytical concepts in QM, that concept of information plays a key role in QM.[2] For instance, those concepts account for the basic difference between von Neumann's Type I and Type II processes and that elucidate both the nature of distinction-making measurement at the quantum level and the no-distinctions unitary evolution of isolated systems. The analysis of von Neumann's Type I and II processes in terms of distinctions and indistinctions matches perfectly with the Feynman's rules analysis in terms of distinguishability and indistinguishability. Thus this analysis of QM starts with the partition logical notions of distinctions and indistinctions, quantitatively developed as logical entropy (information-as-distinctions), and develops those themes, expressed in the vocabularies of identity & difference, indistinctions & distinctions, equivalences & inequivalences, or indistinguishability & distinguishability, which are the themes that go all the way through to von Neumann's two processes and Feynman's rules. It is fitting that those fundamental logical identity & difference concepts are basic to our most fundamental physical theory.[3]

We have systematically constructed at the set level a highly schematic skeletal model using the lattice of set partitions that prefigures quantum states and processes:

- set partitions (as equivalence relations) with point probabilities prefiguring density matrices for quantum states;
- set partitions as the inverse image of real-valued numerical attributes prefiguring the direct-sum decompositions of the eigenspaces of the Hermitian operators as quantum observables;
- the join of set partitions prefiguring the projective measurement of a state represented by a density matrix by an observable; and
- the join of several set partitions on the same set to reach the discrete partition with blocks of cardinality one (CSCA) prefiguring the join of the eigenspace direct-sum decompositions of commuting observables to reach the direct-sum decomposition of subspaces of dimension one (CSCO).

The Yoga of Linearization was used not only to transfer set concepts into the corresponding vector-space concepts, but also to transfer the quantitative notion of

---

[2] Perhaps this is how to better understand John A. Wheeler's speculations about "it from bit." (Wheeler 1999) Otherwise, there has certainly been much speculation about "information" and quantum mechanics. And then Rota's equivalence, information is to partitions as probability is to subsets, hints that partitions may play a role in understanding QM.

[3] The identity & difference concepts are called "logical" instead of the wide-ranging term "metaphysical" since the concepts are grounded in the logic of partitions.

information-as-dits, i.e., logical entropy, into the quantum notion of information-as-qudits, i.e., quantum logical entropy. In the literature, much emphasis has been placed on observables as operators just as partitional mathematics is often formulated in terms of numerical attributes, e.g., random variables. But the logical information is not in the specific numerical values or eigenvalues but when the values are identical or different, and that is the information preserved in the inverse-image partition of a numerical attribute and in the vector-space partition (direct-sum decomposition or DSDs of eigenspaces) given by an observable operator.

In our argument for the origins in partitional mathematics, we have taken the mathematics of QM to be sufficiently represented by the following topics.

5.2   Commuting, Non-commuting, and Conjugate Observables

The notions of commuting, non-commuting, and conjugate observables are usually defined in terms of operators which seem to have no relation to partitional mathematics. But we have argued that the important information for observable operators was captured in the DSD of the eigenspaces–which are the linearized versions of the blocks of a set partition that is the inverse-image of a numerical attribute.

Moreover, we saw that the status of commuting, non-commuting, and even conjugate observables can all be defined in terms of the join-like operation on the eigenspace DSDs of the observables. Thus those properties of operators are really properties of the vector-space partitions (the direct-sum decompositions) whose blocks are the eigenspaces.

One side benefit of the Yoga of Linearization is the development of the intermediate state of vectors-as-subsets in $\mathbb{Z}_2^n \cong \wp(U)$. Running the Yoga in reverse and stopping at this intermediate stage yields a pedagogical (or 'toy') model of quantum mechanics over sets, QM/sets (Ellerman 2017). Then an example of conjugate direct-sum decompositions can be easily developed in any $\mathbb{Z}_2^n$ for even $n > 2$, which shows that conjugacy is not specific to QM concepts.

5.3   Time-Dependent Schrödinger Equation and vN Type 2 Processes

The Schrödinger equation seems, at first, to have little connection to the mathematics of partitions. But we have seen that Von Neumann's notion of Type I processes (projective measurements) was characterized as an interaction that forces distinctions between the eigenstates in an evolving superposition. Hence the natural characterization of the Type II processes would be ones that make no distinctions. The degree of distinctness or indistinctness between two quantum states is measured by their overlap, i.e., their inner product. Hence the natural characterization of the Type II processes are one's that preserve that degree of distinctness between quantum states, i.e., processes that preserve inner products–which are the unitary transformations. The connection to Schrödinger's equation comes out of the mathematics of Stone's Theorem (Stone 1932) which shows that unitary transformations can be represented by the solutions to Schrödinger's equation.

5.4 Measurement and the Collapse Postulate

We have seen how measurement at both the set level and quantum level can be described by a Lüders mixture operation, e.g., applied to incidence matrices and density matrices (representing equivalence relations) at the set level. But the outcome of that calculation is only a more definite mixed state, not a specific outcome. The collapse postulate specifies that the outcome of a non-degenerate or maximal measurement will be an eigenstate (or eigenspace in the degenerate case) with a certain eigenvalue with the probabilities supplied by Born's Rule.

In classical physics, we are accustomed to equations that will describe the continuous transition from one definite state to another. It is this assumption from classical fully-definite physics that calls for some equations to describe the quantum jump from a superposition to a specific eigenstate in the postulated collapse. In partition math, the Principle of Identity of Indistinguishables characterizes the classicality of the discrete partition. But less refined partitions involve non-singleton blocks that represent ontic indefinite states that violate the idea of fully-definite reality at the quantum level. Even at the set level, a choice function applied to non-singleton sets already shows an indeterministic jump. Nature can make (quantum) jumps after all. In QM, the whole classical notion of the trajectory of a particle disappears.

5.5 Particle Indistinguishability

The quantum statistics based on the indistinguishability of particles of the same type gives perhaps the best argument that reality at the quantum level cannot be fully definite, i.e., cannot be definite all the way down. Otherwise there would be further properties that would make an ontic distinction between numerically distinct bosons–which would yield the Maxwell-Boltzmann statistics.

Given a maximal CSCO-defined description of a quantum state, one might expect two types of particles: those for which the maximal description is sufficient to only allow one or no particle in that state, or where there can be many particles in a maximally circumscribed state. The enumerative combinatorics behind the statistics in the two cases can be computed using the standard balls-in-boxes calculations which are stated in terms of our analytical concepts.

In the one-or-none case, the count is of one-to-one (i.e., distinction-preserving) functions from the set of indistinguishable balls (particles) to distinguishable boxes (states) so that two balls could not be assigned to the same box. The balls are still indistinguishable, but the maximal state description suffices to limit one ball in that the state. That yields the Fermi-Dirac statistics. In the other case, the count is of arbitrary functions from the set of indistinguishable balls to the distinguishable boxes, and that yields the Bose-Einstein statistics (Feller 1968, p. 40).

The Maxwell-Boltzmann statistics are obtained when both the balls and boxes are distinguishable and the distribution is by arbitrary functions. Here again we see that the key analytical concepts are distinctions versus indistinctions or distinguishability versus indistinguishability. As in the partition lattice, in the case of full distinguishability, we of course have the classical case, MB statistics.

5.6 Projective Measurement and the Lüders Mixture Operation

The basic logical partition operation that creates distinctions is the join operation. Running the Yoga in reverse gear, it was shown that the Lüders mixture operation, which represents projective measurement at the quantum level, can be applied at the set level using incidence matrices or density matrices. Moreover, we have seen how the classical and quantum notions of information as dits or qudits can be used to measure the distinctions created by the process of measurement by the non-zero off-diagonal elements in the incidence or density matrices that are zeroed in the process. At the set level, when the density matrix represents a partition π and the numerical attribute being 'measured' gives an inverse-image partition σ, then the Lüder's mixture operation just gives the density matrix for the partition join $\pi \vee \sigma$, i.e., the density matrix representing the intersection indit $(\pi) \cap$ indit $(\sigma) =$ indit $(\pi \vee \sigma)$, of the corresponding equivalence relations.

These results directly transfer to the quantum level where the DSDs of the observable being measured replaces the inverse-image partition σ of the numerical attribute and the density matrix of the quantum state replaces the partition π with its point probabilities. These intimate connections are surely not accidental; the quantum mathematics of measurement is the linearized version of the 'measurements' at the level of partitional mathematics. This is a case where the relevant QM mathematics (i.e., quantum logical entropy) had to be developed to demonstrate our thesis.

This analysis of measurement based on making distinctions also seems to be overlooked by so much of the literature that focuses on the "shifty split" (Bell 1990) between microscopic and macroscopic reality, and on the notion of "decoherence" based on interactions with a macroscopic apparatus or environment (Zurek 2003). But we noted that the Feynman rules are careful to give an analysis of measurement (adding probabilities) versus non-measurement (adding amplitudes) based on distinguishability at a completely quantum level (Feynman et al. 1965). Interactions with macroscopic apparatuses for the benefit of human observation have no role in quantum theory.

5.7   Group Representation Theory

The basic set-level mathematical concept to represent distinctions and indistinction is the notion of a partition or equivalence relation or quotient set. The logic of partitions defines logical operations such as the join on given partitions. But how are partitions or equivalence relations "given" in Nature? That is how the concepts of a group and group representation enter the picture. A set representation of a group is an active or dynamic way to define an equivalence relation (the orbit partition) on a set and a vector space representation is similarly a way to define a direct-sum decomposition (of irreducible subspaces) of a vector space. Thus group representations are part of the partitional mathematics that lifts from the set level to the level of vector spaces by the Yoga of Linearization.

We saw previously that at the set level, the distinction-making operation was the partition join that prefigured projective measurement in QM. The maximum of distinctions is the discrete partition with all the elements of the universe set are fully distinguished as in classical reality. In the case of group representation theory, the group representation on a set is a 'dynamic' way to define an equivalence relation (orbit partition) on the set, so the distinction-making operation, called "symmetry-breaking," (Collier 1996; Muller 2007) is moving to the representation of a subgroup which would make fewer symmetries or indistinctions and whose orbit partition would be a refinement of the previous orbit partition. And again, the maximum of distinctions or symmetry-breaking is achieved by moving to the minimum subgroup, i.e., the identity subgroup, whose orbit partition is the discrete partition.

There are symmetries in fully-definite reality of classical physics so groups will play an important role, e.g., the Noether Theorems. But group representations play a more fundamental role in quantum mechanics as should be expected from their role in the mathematics of partitions and equivalence relations. This is in part due to the objectively indefinite states given by the superposition principle, but, at a more basic level, it is due to the irreps (the restriction of representations to the irreducible subspaces) of certain group representations defining the elementary particles at the quantum level.

> The reason is fundamentally, that the variety of states is much greater in quantum theory than in classical physics and that there is, on the other hand, the principle of superposition to provide a structure for the greatly increased manifold of quantum mechanical states. The principle of superposition renders possible the definition of the states the transformation properties of which are particularly simple. It can in fact be shown that every state of any quantum mechanical system, no matter what type of interactions are present, can be considered as a superposition of states of elementary systems. The elementary systems correspond mathematically to irreducible representations of the Lorentz group and as such can be enumerated. (Wigner 1967, p. 8)

5.8 Final Remarks

We have taken the mathematics of quantum mechanics to be adequately represented by:

- density matrices representing quantum states, quantum partitions (DSDs) representing observables, and projective measurement (vN Type I process) represented by the Lüders mixture operation;
- commuting, non-commuting, and conjugate observables;
- evolution by the Schrödinger equation (vN Type 2 process);
- measurement and the collapse postulate;
- quantum statistics for indistinguishable particles; and
- group representation theory in quantum mechanics.

And we have argued, in each case, that the mathematics of QM is the linearized to (Hilbert) vector space version of the mathematics of partitions. Again and again, the connection between the set-based math of partitions (or equivalence relations) and the Hilbert space math of QM has been demonstrated. These connections may be avoided through arrogance or ignorance, but that whole translation dictionary is surely not just a coincidence or an accident. The lattice of partitions, used as a skeletal representation of a particle or particles and their states, is not some jury-rigged structure constructed just to represent QM math.

What does that tell us? The QM formalism is the Hilbert space version of the mathematics used to describe indefiniteness, and that tells us that the unintuitive reality that QM so successfully describes is a reality of objective indefiniteness unlike the fully-definite reality presupposed by classical physics. The partition lattice provides a skeletal model of the pure, mixed, and classical states. In that sense, partitions provide the *skeleton key* to unlock the mathematics of QM.

- The key analytical concepts were the notions of definiteness versus indefiniteness, distinctions versus indistinctions, and distinguishability versus indistinguishability, e.g., to characterize both the vN Type I and Type II processes.
- Both the key tools in QM math, observables and quantum states, arise from partitions. The difference is the extra structure associated with the partitions:

  - Partitions that are inverse-images of numerical attributes prefigure the DSDs of observable operators in QM math; and
  - Partitions (or the corresponding equivalence relations) with point probabilities prefigure the density matrices in QM math.

- That both observables and quantum states arise from partitions is to be expected since they need to combine in projective measurement that applies an observable's DSD to a density matrix (Lüders mixture operation) which is the vector space correlate of the join of the two partitions playing those two roles.
- In addition to the quantum jumps between different levels of indefiniteness (Type I processes), there is the distinction-preserving process of unitary evolution which,

unlike classical definite-to-definite processes, can operate at levels of indefiniteness and thus show the characteristic quantum interference effects.

Furthermore, the *anshaulich* lattice of partitions skeletal model makes sense out of the fact that there are two fundamental process in QM and only one in classical mechanics. The top of the lattice, the discrete partition, represents classical definite states so the Type I vertical change from indefinite to more definite is no longer possible, and thus in the classical case, there is only the horizontal 'Type II evolution' of definite states to definite states.

The 'standard' reality-oriented interpretations of QM (e.g., Bohmian mechanics, spontaneous collapse, many-worlds,) essentially ignore those (partition-based) analytical concepts. The partition math approach of showing the set-level origins of the mathematics of QM takes the formalism as being complete–without any addition of other variables ('pilot waves'), other equations (GRW), or other-worldly interpretations of unitary evolution and measurement.

> If our best-confirmed theory of the physical world, taken literally, represents electrons as lacking determinate positions, then we should accept that there is indeterminacy in the world, rather than claiming that the theory is incomplete just to preserve our classical intuitions. (Lewis 2016, p. 78)

The partition approach to understanding of QM can be seen as an explication of Shimony's Literal Interpretation.

> These statements ... may collectively be called "the Literal Interpretation" of quantum mechanics. This is the interpretation resulting from taking the formalism of quantum mechanics literally, as giving a representation of physical properties themselves, rather than of human knowledge of them, and by taking this representation to be complete. (Shimony 1999, pp. 6–7)

By seeing where the QM mathematical formalism "comes from," we can focus on the key concepts (distinctions and indistinctions), machinery, and operations of QM mathematics–as well as find some skeletal imagery (partition lattice), albeit simplified, for the objectively indefinite reality at the quantum level.

> The conceptual elements of quantum theory that now underlie our picture of the physical world include objective chance, quantum interference, and the objective indefiniteness of dynamical quantities. Quantum interference, which is directly observable, was readily absorbed by the physics community. Objective chance and indefiniteness, being of more philosophical significance, gained acceptance only after much debate and conceptual analysis, when it was recognized that observed phenomena are better understood through these notions than through older ones or hidden variables. (Jaeger 2014, p. vii)

In short, the fact that the mathematics of QM is the Hilbert space version of the mathematics of set partitions (or equivalence relations) provides the mathematical basis to interpret QM as describing a reality of objective indefiniteness. That is the *Objective Indefiniteness Interpretation* of quantum mechanics (Ellerman 2022, 2024).

References

Bell JS (1990) Against "Measurement". In: Miller AI (ed) Sixty-Two years of uncertainty. Plenum Press, New York, pp 17–31

Collier J (1996) Information originates in symmetry breaking. Symmetry: Sci Culture 7:247–256

Ellerman D (2017) Quantum mechanics over sets: a pedagogical model with non-commutative finite probability theory as its quantum probability calculus. Synthese 194:4863–4896. https://doi.org/10.1007/s11229-016-1175-0

Ellerman D (2021) New foundations for information theory: logical entropy and Shannon entropy. Springer Nature, Cham, Switzerland

Ellerman D (2022) Follow the math!: the mathematics of quantum mechanics as the mathematics of set partitions linearized to (Hilbert) vector spaces. Found Phys 52. https://doi.org/10.1007/s10701-022-00608-3

Ellerman D (2024) A new logic, a new information measure, and a new information-based approach to interpreting quantum mechanics. Entropy Spec Issue: Inf-Theor Concepts Phys 26. https://doi.org/10.3390/e26020169

Feller W (1968) An introduction to probability theory and its applications, vol 1, 3rd ed. Wiley, New York

Feynman RP, Leighton RB, Sands M (1965) The Feynman lectures on physics: quantum mechanics, vol III. Addison-Wesley, Reading, MA

Heisenberg W (1962) Physics and philosophy: the revolution in modern science. Harper Torchbooks, New York

Jaeger G (2014) Quantum objects: non-local correlation. Springer, Heidelberg

Lewis PJ (2016) Quantum ontology: a guide to the metaphysics of quantum mechanics. Oxford University Press, New York

Muller SJ (2007) Asymmetry: the foundation of information. Springer, Berlin

Shimony A (1993) Search for a naturalistic world view: vol II Natural science and metaphysics. Cambridge University Press, New York

Shimony A (1999) Philosophical and experimental perspectives on quantum physics. Epistemological and experimental perspectives on quantum physics: Vienna Circle Institute yearbook 7. Springer Science+Business Media, Dordrecht, pp 1–18

Stone MH (1932) On one-parameter unitary groups in Hilbert Space. Ann Math 33:643–648

Weyl H (1949) Philosophy of mathematics and natural science. Princeton University Press, Princeton

Wheeler JA (1999) Information, physics, quantum: the search for links. In: Hey AJG (ed) Feynman on computation. Perseus Books, Reading, MA, pp 309–336

Wigner EP (1967) Invariance in physical theory. Symmetries and reflections: scientific essays of Eugene P. Wigner. Ox Bow Press, Woodbridge CN, pp 3–13

Zurek WH (2003) Decoherence, einselection, and the quantum origins of the classical. Rev Mod Phys 75:715–775

Index

© The Author(s), under exclusive license to Springer Nature Switzerland AG 2024   115
D. Ellerman, *Partitions, Objective Indefiniteness, and Quantum Reality*,
SpringerBriefs in Philosophy, https://doi.org/10.1007/978-3-031-61786-7